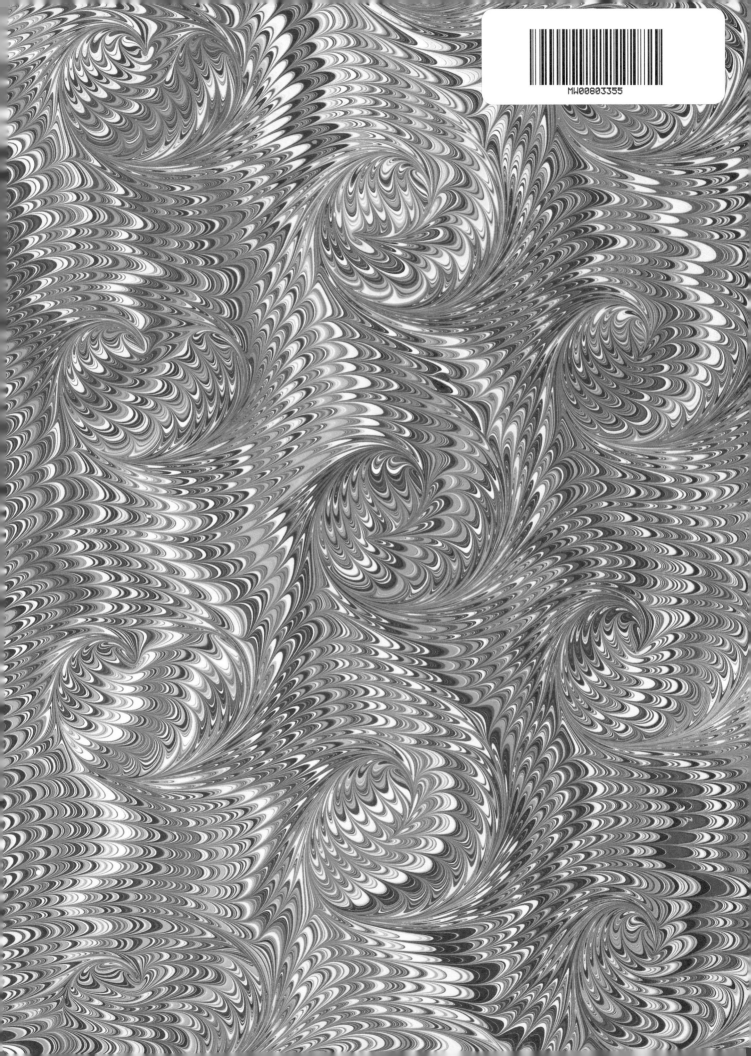

German Jet Aircraft
1939-1945

German Jet Aircraft
1939-1945

Hans-Peter Diedrich

Schiffer Military History
Atglen, PA

Book Design by Ian Robertson.
Translated from the German by David Johnston.

Copyright © 2000 by Schiffer Publishing, Ltd.
Library of Congress Catalog Number: 00-105623

Printed in China.
ISBN: 0-7643-1230-8

This book was originally published under the title,
Die deutschen Strahlflugzeuge bis 1945
by Aviatic Verlag GmbH

We are interested in hearing from authors with book ideas on related topics.

Published by Schiffer Publishing Ltd.
4880 Lower Valley Road
Atglen, PA 19310
Phone: (610) 593-1777
FAX: (610) 593-2002
E-mail: Schifferbk@aol.com.
Visit our web site at: www.schifferbooks.com
Please write for a free catalog.
This book may be purchased from the publisher.
Please include $3.95 postage.
Try your bookstore first.

In Europe, Schiffer books are distributed by:
Bushwood Books
6 Marksbury Avenue
Kew Gardens
Surrey TW9 4JF
England
Phone: 44 (0) 20 8392-8585
FAX: 44 (0) 20 8392-9876
E-mail: Bushwd@aol.com.
Free postage in the UK. Europe: air mail at cost.
Try your bookstore first.

Contents

Foreword

The development of turbojet aircraft is inextricably linked to the development of turbojet engines.

The idea of jet propulsion was born in the early days of aviation. Even before the First World War the Frenchman Armengaud and the German Holzwarth were working on gas turbines for aircraft, however, financial considerations forced both to cease their efforts. Holzwarth built the first German gas turbine in 1908.

In 1910 the Romanian Henry Coanda developed a power plant with a shrouded propeller. The resulting compressed air was mixed with fuel and ignited, increasing the amount of thrust produced. Not only did this mark the birth of the concept of the turbofan, but also of the afterburner. Coanda designed an aircraft powered by this propulsion system and displayed it at the Paris Aviation Salon in 1919. Unfortunately, subsequent attempts to fly the machine resulted in the wood and fabric biplane going up in flames. He, too, was forced to cease his activities because of a lack of funds.

In the Soviet Union scientist V. I. Basarov applied for a patent for a gas turbine in 1923. The poor economic conditions which were a consequence of the October Revolution of 1917 doomed his work to failure. In 1936 engine designer A.L. Lyulka began work on a turbofan engine for aircraft which was supposed to reach the test stage by autumn 1941. The German invasion of the Soviet Union on 22 June 1941 prevented this, however.

In England, Frank Whittle began work on a gas turbine for aircraft in 1928. In 1929 he presented his work to the British Air Ministry, however, it was rejected. After much effort he found a backer, and in 1936 founded the company Power Jets Ltd, where his position was limited to "honorary chief engineer." In March 1937 the first experimental turbine, the Whittle U-1, was severely damaged when the compressor wheel disintegrated. Finally, on 12 April 1937, the turbine completed a test run. Whittle's progress attracted the attention of the British Air Ministry in 1939, and in June of that year he received a contract to build a turbine for an aircraft. Gloster received a contract (E28/39) to build a turbojet aircraft, which made its first flight on 15 May 1941.

In Germany, scientist Pabst von Ohain began working on a turbojet engine in the early 1930s; his ideas were well-received by Prof. Pohl of the University of Göttingen, and a short time later by aircraft builder Heinkel. On 27 August 1939 Ohain's work reached its culmination when the He 178, powered by a Heinkel HeS 03 engine, took off on its first flight from the Heinkel company airfield at Rostock-Marienehe, launching the turbojet age.

Another pioneer was Dr. Hermann Oestrich, who began investigating turbojet propulsion for aircraft in 1928, first with the DVL and later with Bramo. BMW ultimately profited from his work, producing the BMW 003 turbojet engine.

During this period, work on turbojet propulsion was conducted almost simultaneously in various European states and in the USA, and each believed that it was the first to do so. British and American consternation must have been great when the world's first production turbojet fighter, the Me 262, appeared in the skies over Germany in 1944.

It is worth noting that as early as 4 January 1939 the German air ministry (*Reichsluftfahrtministerium*) issued a specification to the German aviation industry titled "Provisional Technical Guidelines for Fast Fighter Aircraft with Turbojet Propulsion." Among the requirements contained therein was a maximum speed of 900 km/h. This should not conceal the fact, however, that in Germany it was designers Heinkel and Messerschmitt who first recognized the advantages of turbojet propulsion and pursued its development. In spite of lukewarm support from the RLM, Ernst Heinkel personally financed the development of the world's first turbojet aircraft and the first twin-turbojet fighter aircraft. And it was Professor Messerschmitt who, after much effort, produced the world's first production turbojet fighter.

On 20 February 1944 the Allies launched "Big Week," a series of major air strikes intended to bring the German aviation industry to its knees. One of the consequences of these attacks was the formation by the Germans of the so-called "Fighter Staff" (*Jägerstab*) on 1 March 1944. It was formed in response to the threat posed by Allied air attacks,

and in order to avoid protracted bureaucratic channels it was directly subordinated to the head of the Armaments Ministry. In addition, facilities associated with aviation were to be further spread out, destroyed factories rebuilt, and underground or specially protected production facilities constructed.

All of these measures led to a real upswing in the production of German turbojet aircraft. The number of aircraft types to be built was also severely restricted. From that point on the aviation industry would concentrate on the Me 262, Ar 234, Ju 287, Do 335, and Ta 152. In spite of loud objections (especially by Messerschmitt) that power plant development had not yet reached the point where it could justify a single-engined aircraft, the He 162 was added to the list in the autumn of 1944. These measures made it possible for the aviation industry to achieve its highest production rates of the entire war in the summer of 1944 in spite of repeated air attacks. Nevertheless, at the beginning of 1945 production collapsed. The Allies had found that while they could not achieve long-lasting effects against protected or hidden production sites (if they were even aware of them), they could cripple the transportation system which linked the various production facilities, especially through the use of fighter-bombers. The production of power plants, in particular, began to suffer, and many completed airframes sat idle without engines.

Turbojet Aircraft

Heinkel He 178

The development history of the He 178 is very closely linked with the development of turbojet engines by Heinkel.

Heinkel immediately recognized the advantages of turbojet propulsion for aircraft after a conversation with Pabst von Ohain on 17 March 1936. He subsequently used private financing to begin development of a turbojet engine for use in aircraft. At the same time he instructed his airframe designers Siegfried and Walter Günther to design and build an aircraft to accept the turbojet engine, the He 178. For this purpose a special development section (*Abteilung Sonderentwicklung II*) was set up in the factory at Rostock-Marienehe; it was kept under the tightest secrecy and completely sealed off from the outside. Even the RLM and its technical office (*C-Amt*) were unaware of Heinkel's activities.

The He 178 was laid out as a cantilever high-wing monoplane with a retractable tailwheel undercarriage operated by compressed air. The main problem was installing the power plant in the fuselage aft of the cockpit, something that had not been done before. The engine was attached to the fuselage by four brackets. The air intake was positioned in the center of the fuselage nose, and air was directed through a flattened tube under the cockpit to the engine. Exhaust gases were directed through a slightly conical tube to the end of the fuselage, where there was an oval turbojet exit with two flow regulator flaps. These flaps could be controlled from the pilot's position. When the aircraft was station-

ary, the flaps were fully opened, while in high-speed flight they were closed, reducing the size of the jet orifice and achieving maximum exhaust gas flow speed.

The first test run by the first HeS 2 engine was carried out in March 1937, however, the He 178 airframe was not completed for another two years. In the intervening period flight trials were conducted with the HeS 3 power plant (450 kg static thrust) mounted beneath the fuselage of an He 118, with encouraging results. Tests were limited, however, for after one landing residual fuel caused the HeS 3 A power plant to catch fire, and it was completely destroyed. The slightly more powerful HeS 3 B engine (500 kg static thrust) was installed in the He 178.

On the early morning of 27 August 1939, Erich Warsitz took the He 178 up on its maiden flight. He described this flight several years later at a press conference marking the twentieth anniversary of the He 178's first flight:

"It was between 3 and 4 in the morning on Sunday, the 27th of August, 1939,

The Heinkel He 178, the world's first jet aircraft, prior to takeoff on 27 August 1939.

The second prototype of the He 178, which never flew (above and below).

just a few days before the outbreak of the war. The machine was towed out to take-off position. Since Heinkel had built this machine entirely on its own and without informing the RLM, we had not been allowed to be seen with it, even on the factory grounds, for otherwise secrecy would have been broken. Heinkel wanted to present the RLM with a complete aircraft. As a result, I was forced to dispense with taxiing trials, which were normally a requirement. I again checked the controls for freedom of movement,

checked the running engine at different rpm settings, the pump pressures, temperatures, and much more, and then gave the fitters the signal to close the canopy.

After a takeoff roll of about 300 meters I accelerated quickly. It was wonderfully simple to maintain direction with the brakes, and then it lifted off...

I repeatedly tried to retract the undercarriage, but something was wrong. The main thing was, it was flying. The airspeed indicator showed 600 km/h,

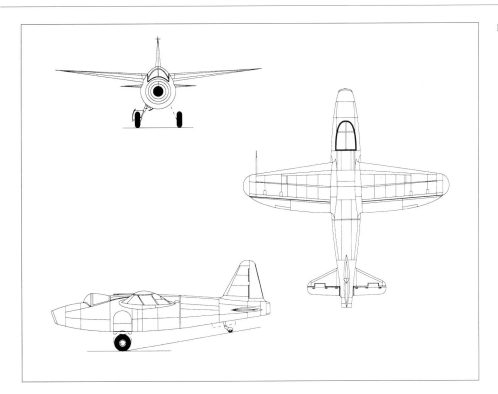

and Herr Schwärzler (one of the designers) had told me that under no circumstances was I to fly faster. I was therefore forced to reduce power.

I had been airborne for six minutes and had to prepare to land. The turbine reacted obediently to my throttle lever, although one fuel pump had failed. This was indicated by my instruments and later confirmed by examination of the engine.

I leveled off rather too high, but my fuel state prevented me from making another circuit. For better or worse I had to sideslip, for otherwise I would have rolled into the Warnow. I was less than enthusiastic about the idea of a sideslip in a new machine, especially so close to the ground. The aircraft touched down smoothly without bouncing, rolled out without any tendency to swing, and I brought it to a stop right in front of Dr. Heinkel and his group."

Not until 1 November 1939 did representatives of the RLM (Udet, Milch, and Lucht) go to Heinkel to see the He 178 fly, however, they showed little or no interest in the aircraft and its new power plant technology.

Two prototypes were built, the He 178 V1 and V2. The second prototype differed from the first in having an increased wingspan (8.60 m). The V2 was held in reserve to take the place of the V1 in the event of a crash during the flight test program, however, it never flew. Heinkel painstakingly reviewed the results of the He 178 test program and used the data in the development of the He 280.

Heinkel He 280

The good results achieved with the He 178 prompted Heinkel to develop a fully-equipped turbojet fighter, the He 280.

Installation of the power plant in the He 178's fuselage and the associated ducting had led to a thrust loss of about 15%. For this reason, and because the reliability of the power plants still left something to be desired, the designers of the He 280 decided to install two turbojet engines in pods beneath the wings, a solution that was also adopted for the Me 262.

He 280 V1 with HeS 8 engines.

At the same time, development work began on the more powerful HeS 8 engine. Intended as it was for installation beneath the wings, the engine had to have a significantly reduced diameter compared to the HeS 3.

The He 280 introduced two innovations:

• For the first time in Germany an aircraft was equipped with a nosewheel undercarriage. The nosewheel had previously been tested on an He 100.

• For the first time in the world an ejection seat (compressed air) was installed in an aircraft.

The airframe of the He 280 V1 was completed on 20 September 1940. As power plants were not yet available, it was decided to conduct unpowered trials with engine mockups in place in order to conduct handling trials. On 22 September 1940 the He 280 V1 was towed by an He 111 B to an altitude of 4,000 meters and released. The flight

He 280 Technical Description

Purpose:	Twin-engined single-seat fighter
Crew:	One pilot in an enclosed cockpit with bubble canopy
	Heinkel compressed-air ejection seat
Wing:	Cantilever mid-wing monoplane, single-spar all-metal wing with straight leading edge and elliptical trailing edge.
	Landing flaps in center-section.
	Ailerons with trim tabs.
Tail:	Cantilever all-metal tailplane.
	High-placed horizontal tail.
	Twin fins and rudders.
Undercarriage:	Retractable tricycle undercarriage.
	Mainwheels retract inwards into the wing center-section, nosewheel retracts rearward into fuselage nose.
Power Plants:	Two Heinkel HeS 8 A turbojets each producing 700 kg of static thrust
Armament:	Three MG 151/20 cannon in fuselage nose

Heinkel He 280 V3

and subsequent landing were uneventful. The HeS 8 power plants became available at the beginning of 1941 and were installed in the He 280 V2, which was completed at the end of March. The turbojet engines had previously been flight-tested beneath an He 111. On 2 April 1941, with Heinkel company pilot Fritz Schäfer at the controls, the He 280 took off from Rostock-Marienehe on its maiden flight. The engines had a project-ed life of just one hour, and there were cooling problems, as well; consequently, the first flight was made without engine cowlings. Furthermore, the engines were restricted to 13,000 rpm; the pilot had to regulate engine speed manually, since there were no engine controls in the present-day sense. Just three days later, on 5 April 1941, the He 280 was demonstrated to representatives of the RLM, including *General* Udet, by Erich Warsitz. During the course of the flight Warsitz achieved a speed of 776 km/h at an altitude of 6,100 meters. Contemporary piston-engined fighters were capable of speeds in the order of 560 km/h.

A total of nine prototypes of the He 280 were built. On 27 March 1943 the program was halted by the RLM in favor of the Me 262, and the He 280 never entered production.

He 280 on its maiden flight. Note that engine cowlings are not fitted.

The prototypes were used for other purposes, including the testing of various engines. A developed version of the fighter was supposed to be powered by the HeS 011 engine, however, it was still in development when the war ended.

On 18 November 1942 work began to convert the He 280 V1 to take two Argus As 014 pulse-jets. It completed its first flight in this configuration on 13 January 1943. Because of the performance characteristics of the Argus engines, the He 280 was carried on top of an He 111 to an altitude of 2,400 meters

List of He 280 Prototypes	Werk-Nr.	Code	First Flight	Remarks
V1	280 000 001	DL + AS	22/09/1940	After first unpowered flight equipped with HeS 8 A turbojets delivered to the Rechlin Test Station. There the engines were replaced with four As 014 pulse-jets. On its first flight on 13/01/1942 it was towed to altitude by an He 111 but then crashed due to icing. Pilot Schenk of the Argus company ejected to safety.
V2	280 000 002	GJ + CA	30/03/41	First equipped with two HeS 8 A engines, then Jumo 004s. Experimental armament of three MG 151/20. Almost completely destroyed in crash-landing on 26/04/1942.
V3	280 000 003	GJ + CB	summer 1942	Ready to fly on 15/01/1943, served mainly for HeS 8 A testing.
V4	280 000 004			First equipped with two BMW 003 turbojets, later converted to four Argus As 014 pulse-jets at the Rechlin Test Station.
V5	280 000 005			Engine trials with HeS 08 A and BMW 003, reached 820 km/h in a dive.
V6	280 000 006	NU + EA		Trials with BMW 003 engines.
V7	280 000 007	NU + EB	19/04/1943	Transferred to the DFS at Ainring for pressure measurements on wing, engines removed.
V8	280 000 008	NU + EC	19/07/1943	Transferred to the DFS at Ainring for testing of a vee-tail and other special installations.
V9	280 000 009			Engine trials with BMW 003 E, from August 1943.

and released. Later in the flight Argus company pilot Schenk was forced to eject and the V1 was completely destroyed. This was probably the first time in history that a pilot's life was saved by an ejection seat. The Heinkel system, which was developed in the period from 15 June to 17 November 1940, was operated by compressed air, which fired the seat approximately 5.70 meters out of the aircraft. The seat itself weighed just 40 kg.

The He 280 V2 was converted to take two Jumo 004 engines and completed its first flight in this configuration on 16 March 1943. In the course of subsequent flights the V2 reached speeds of 800 km/h. Some instability about the vertical axis was detected, requiring modifications. This instability was probably one of the reasons why the program was canceled.

The He 280 V4 to V6 were supposed to serve as flying test-beds in the BMW 003 development program, however, it is unlikely that they ever flew in that role.

The V7 completed its first, unpowered flight on 19 April 1943, and was assigned to the DFS in Ainring, near Linz, as an unpowered research aircraft.

The He 280 V8 made its first flight on 19 July 1943 powered by two HeS 8a turbojet engines, and was likewise assigned to DFS in Ainring. The V8 was later fitted with a vee-tail as part of the He 162 test program.

Heinkel He 162 *Volksjäger*

Reacting to the growing weight of the Allied bombing offensive, on 8 September 1944 the RLM issued a requirement for an interceptor fighter to Arado, Blohm & Voss, Junkers, and Heinkel. Quantity production was supposed to begin in January 1945.

Göring invited the participating companies to a *Volksjäger* conference to be held on 23 September 1944. There they were to present their designs. On that same day a decision was made in fa-

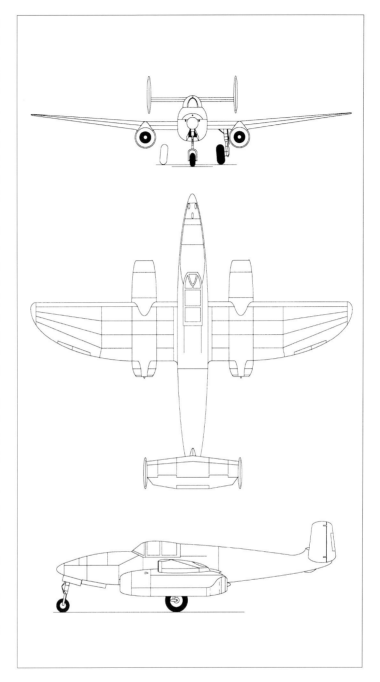

Heinkel He 280

vor of the He 162, which was based on Heinkel's *Projekt P.1073*.

The Heinkel P.1073 possessed the following characteristics:
- high-speed fighter
- critical mach number higher than that of the Me 262
- 1 x HeS 011A (max. speed 1 010 km/h) or 1 x Jumo 004C turbojet engine (max. speed 940 km/h)
- armament of 3 x MG 151/20 cannon
- was to replace the Me 262

With only a short time available for development, Heinkel opted to mount the power plant on top of the fuselage and produce as many components as possible from wood. The aircraft would, however, be equipped with the ejection seat developed by Heinkel. By this time a tricycle undercarriage was considered obligatory for turbojet-powered aircraft.

Heinkel began designing the aircraft the very day after the meeting. Building specifications were completed on 15 October 1944, and the complete set of drawings by 5 November. Construction of the first prototype began immediately, and on 6 December 1944 Ing. Peter took the He 162 V1 up on its first flight.

Then, on 10 December 1944, with the aircraft in a 10-degree dive at a speed of 700 km/h, the plywood leading edge of the starboard wing separated, the wooden skinning literally peeled off, and finally the aileron separated from the wing. The aircraft crashed from a height of 100 meters. The cause was poor gluing of the leading edge.

In spite of this setback the He 162 was a firm part of Production Plan 227 (dated 15/12/1944). This plan called for production of the Ju 88, Ju 388, He 219, and Do 335 to be scaled back in favor of the Me 262 and the He 162. Beginning in July 1945, 750 He 162s were to be delivered per month.

On 22 December 1944 the V2 completed its first flight without incident. At the end of December 1944 the system for designating prototype aircraft was changed, with the adoption of the letter M in place of the V previously used. Two further prototypes, the M3 (V3) and M4

(V4), were completed by 16 January 1945, both with extended wings. While the prototypes were engaged in flight trials, production of the He 162 began at Vienna-Schwechat and Rostock. However, the components delivered by the decentralized subcontractors were of such poor quality that Heinkel had to set up its own department to rework these components. As well, technical changes which had to be incorporated into the production line resulted in delays in delivery of the first production machine. Stability about the longitudinal and normal axes had to be improved, and a cure found for the aircraft's strong tendency to side-slip. On the advice of Prof. Lippisch, new wingtips canted down at 45 degrees were adopted, the so-called "Lippisch Ears." Three production sites were selected for final assembly: Hei-

The RLM Issued the following Requirements:

Maximum speed:	750 km/h
Power plant:	one BMW 003
Armament:	two MG 151/20 or two MK 108
Endurance:	minimum 20 min. at low altitude
Gross weight:	less than 2 000 kg

simplest airframe structure
no curtailment of the Me 262 and Ar 234 programs

nkel-North (Werk-Nr. 120 xxx), Heinkel-South (Werk.Nr. 220 xxx), and Junkers at Bernburg (Werk.Nr. 300 xxx).

Finally, on 28 January 1945, the M5 was completed and was declared the first He 162 A-01 production machine. This step was taken even though there were still unsolved stability problems.

For this reason the fuselage of aircraft M25 was lengthened. Another problem that was never solved was fuel delivery to the engine. Under high g-loads the fuel supply was interrupted and the fuel pumps sucked air, which made restarting the engine even more difficult. All of this indicated that the He 162 was far

He 162 Technical Description

Purpose:	Single-seat interceptor fighter
Crew:	One pilot, clear-view hood, ejection seat
Wing:	Cantilever shoulder-wing monoplane of wooden construction. One-piece trapezoid-shaped wing with straight leading edge and downward-canted wingtips. Specially impregnated and sealed for acceptance of fuel.
Fuselage:	Fuselage nose of wooden construction, fuselage center- and rear sections of metal construction.
Undercarriage:	Retractable tricycle undercarriage, rearwards into fuselage.
Power plant:	One BMW 003 E producing 800 kg of static thrust mounted on fuselage spine.
Armament:	Two MG 151/20 in lower fuselage.

Heinkel He 162 on display in Hyde Park in London on Battle of Britain Day, September 1945.

from being a "Peoples Fighter." Only an experienced pilot could fly the machine and exploit its potential. Production aircraft had problems reaching the required maximum speed, however, this was more a problem with the engine than the airframe.

Production of the He 162 proceeded slowly, nevertheless, on 6 February 1945 I./JG 1 began converting to the type. By 7 February 1945 the *Gruppe* had handed over its last Fw 190 on the island of Usedom. Instruction of personnel lasted until 31 March 1945, partly on account of the numerous transfers which the unit had to complete.

The Fw 190 and Bf 109 pilots received a hurried conversion course at Parchim. There were no training aircraft. Training manuals were used to familiarize the pilots with the aircraft, which they had to master after one flight. Meanwhile, II./JG 1 had also converted to the He 162. At the beginning of May 1945 both *Gruppen* moved to Leck airfield.

The first official mission by the He 162 took place against British tactical aircraft on 26 April 1945. On 6 May British troops occupied Leck airfield, ending the He 162's brief operational history. Unavoidable bottlenecks in the production of the BMW 003 power plant (just 100 of the specified 5,000 units were completed in March 1945, and 60% of production was reserved for the Ar 234 and 40% for the He 162, while all Jumo 004 production was set aside for the Me 262) resulted in two sub-variants of the He 162 A-10 with two Argus As 014 pulse-jets on top of the fuselage, and the He 162 A-11 with a more powerful As 044 pulse jet on the fuselage spine. Construction of the He 162 A-10 was completed in March 1945, and it made its first flight. The He 162 A-11 was not finished, since the Argus As 044 was never delivered. Both variants featured a fuselage lengthened by 9.20 meters in order to accommodate an additional fuel tank, and the canted wingtips were dropped. Because of the operating characteristics of the pulse jet engines, both variants were supposed to be launched from a catapult or equipped with takeoff assist rockets.

List of Prototypes

Werk-Nr.		First Flight	Remarks
V1	200 001	06/12/1944	one BMW 003 E turbojet. Crashed during second flight on 10/12/1944.
V2	200 002	22/12/1944	one BMW 003 E turbojet, later equipped with two MK 108 cannon.
M3	200 003	16/01/1945	one BMW 003 E turbojet, from 18/02/45 served to test enlarged fins and rudders.
M4	200 004	16/01/1945	one BMW 003 E turbojet, new strengthened wing and achieved 700 km/h
M5	200 005		vibration tests from 24/12/1944
M6	200 006	23/01/1945	one BMW 003 E turbojet, two MK 108 cannon, later replaced by two MG 151/20. Crashed on 04/02/1945 because of jammed rudder.
M7	200 007	Feb. 1945	one BMW 003 E turbojet, braking parachute, two MK 108 cannon later replaced by two MG 151/20.
M8	200 008	Feb. 1945	one BMW 003 E turbojet, two MK 108 cannon later replaced by two MG 151/20.
M9	200 009		similar to M8 but with two MK 108
M10	200 010		as M9
M11	220 017		one Jumo 004, never flown
M12	220 018		one Jumo 004, never flown, but possibly tested with one BMW 003 E
M14			one HeS 011 (project only)
M15			one HeS 011 (project only)
M16	220 019		two-seat trainer
M17	220 020		two-seat trainer
M18	220 001	24/01/1945	one BMW 003 E turbojet, first aircraft from Hinterbruhl (A-01), used for undercarriage and radio trials.
M19	220 002	28/01/1945	one BMW 003 E turbojet, designation A-02, delivered to 2./JG 1 and destroyed on 14/03/1945 as a result of pilot error.
M20	220 003	10/02/1945	one BMW 003 E turbojet, simplified undercarriage, destroyed on 25/02/45.
M21	220 003	10/02/1945	one BMW 003 E turbojet, weapons tests with two MG 151/20.
M22	220 005	25/02/1945	one BMW 003 E turbojet, revised wing
M23	220 006		one BMW 003 E turbojet, similar to M22
M24	220 007		
M25	220 008	17/02/1945	one BMW 003 E turbojet, lengthened fuselage, destroyed on 02/03/45.
M26	220 009		one BMW 003 E turbojet, similar to M25
M27	220 010		reserve aircraft
M28	220 011		reserve aircraft
M29	220 012	18/02/1945	one BMW 003 E turbojet
M30	220 013	24/02/1945	one BMW 003 E turbojet, equipped with "Adler" weapons sight.
M31	220 014		destroyed in bombing raid on 13/02/1945.

In spite of technical problems at the start of quantity production and the severe restrictions imposed by the military situation in 1945, approximately 200 He 162s were completed. Of these only a handful saw action, however.

Ernst Heinkel stated rightly that the He 162 could be seen as the last great effort by the German aircraft industry in the field of jet aircraft construction.

Heinkel He 343

In January 1944 Heinkel began work on a turbojet-powered bomber which was supposed to go into production in the summer of 1945 as a competitor to the Ju 287. With such little time available for development, risks associated with the aircraft's handling qualities and operating safety had to be kept as low as possible. Consequently, it was decided

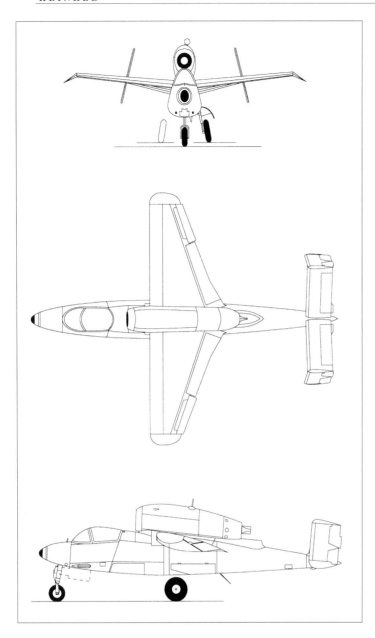

Heinkel He 162 A-1

to base the project on the Heinkel P.1068, a conventional four-jet bomber, and adopt the layout of the Arado Ar 234, which was then achieving favorable results in flight trials. The Ar 234 had reached speeds of up to 1,000 km/h in a dive with no aerodynamic problems. It was decided to increase the size of the Arado wing by a factor of 1.5 and retain its unswept, trapezoidal form. The fuselage had to be enlarged to accept a 2,000 kg bomb load. As well, four Jumo 004 power plants would be installed in nacelles beneath the wings. Maximum allowable takeoff weight would be 16,000 kilograms. Calculated empty weight in this configuration was 5,260 kilograms. The He 343 was to be flown by a crew of two. Design work began in February and was complete by the end of 1944.

Heinkel first presented the project to the RLM at the beginning of 1944 as part of the "16-tonne Jet Bomber" project, but it was not until a second attempt in April 1944 that the RLM ordered the construction of twenty prototypes of the Heinkel bomber, which was designated the He 343. The prototypes were to be assigned *Werknummer* 850 061 to 850 080. Work on the prototypes began immediately in the summer of 1944. The V1, which was to be used for general flight trials with minimal equipment, was supposed to be completed by

15 April 1945. The V2 was to be laid out as a bomber and used for weapons and radio trials. The V3 was to be a prototype for the heavy fighter (*Zerstörer*) variant. Projected completion date was 30 June 1945, and it was to be fitted with twin tailplanes and full armament. Finally, the V4 was conceived as a three-seat bomber to be used as a launch platform for *Fritz-X* and Hs 294 guided weapons. The Emergency Fighter Program resulted in numerous delays in construction of the prototypes, and on 2 March 1945 the program was halted and all finished components were placed in storage.

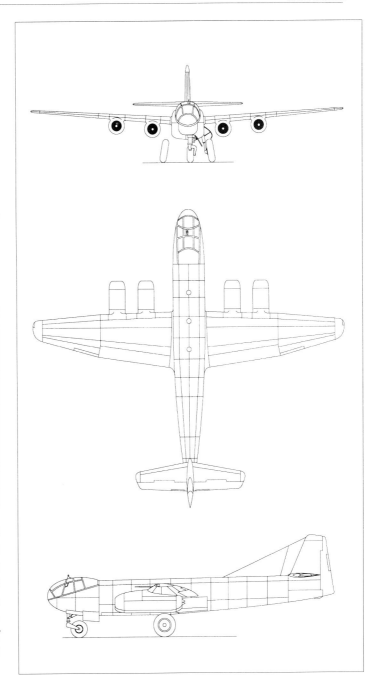

Three Operational Versions of the He 343 Were Planned:

He 343 A-1	Bomber with a bomb load of 3 000 kg (2 000 kg in fuselage and two 500-kg bombs beneath the inner engine nacelles)
He 343 A-2	Reconnaissance aircraft with an additional armored fuel tank in the bomb bay
He 343 A-3	Heavy fighter with an gun pack in the bomb bay (four MK 103 or two MK 103 and two MG 151/20)

The three versions of the He 343 were to be powered by engines in the 900 to 1,300 kg thrust class (Jumo 004B, Jumo 004C, BMW 003C, HeS 011A) and were to be equipped with a fixed rearward-firing defensive armament of two MG 151/20 cannon. Takeoff weights ranged from 16 425 to 19 550 kilograms.

Heinkel He 343

List of Prototypes

	Werk-Nr.	Completion (planned)	Remarks
V1	850 061	15/04/1945	Four Jumo 004 C, basic equipment only, work halted in March 1945
V2	850 062		Four Jumo 004 C, bomber version, to be used for weapons and radio trials
V3	850 063	30/06/1945	Heavy fighter version, twin fins and rudders
V4	850 064	15/08/1945	Three-seat bomber for Fritz X of Hs 294 stand-off weapons.
V5	850 065		Heavy fighter with modified wing for very high altitudes (project).
V6	850 066		Heavy fighter, supposed to receive wooden wing (project).

Messerschmitt

Me 262 *Schwalbe* (Swallow)

The research work which was to result in the Me 262 began on 1 April 1939. On that day the project department of the Messerschmitt AG began work on a jet-propelled fighter under the designation P65. Just three months later, on 7 June 1939, Messerschmitt was able to present project submission P1065 to the RLM.

On 31 January 1940 the RLM issued a contract for the construction of twenty prototypes. The first three aircraft were to be powered by BMW 3302 turbojet engines, which required some modification of the original design:

- triangular fuselage cross-section
- armament of three MK 108 cannon
- swept outer wing panels to compensate for the heavier power plants
- wingspan   12.35 m
- length   10.46 m

As before, the engines were to be buried in the wings. Not until November 1940 were the wings redesigned to accommodate the engines in underwing nacelles, simplifying the design of the main spar.

P1065 Specification

Purpose:	Pursuit fighter
Propulsion:	Two BMW turbojets, each 600 kg of thrust
Wingspan:	9.40 m
Height:	2.80 m
Length:	8.30 m
Wing area:	18.0 m^2
Performance:	
Landing speed at normal weight:	130 km/h
Landing speed at max. weight:	164 km/h
Time to climb to 6 000 m at 100% power:	6.16 min.
Time to climb to 6 000 m at 130% power:	3.98 min.
Takeoff distance to 20 m at 100% power:	800 m
Maximum speed at 3 000 m at 100% power:	840 km/h
Maximum speed at 3 000 m at 130% power:	950 km/h
Landing weight with ammunition and fuel:	3 196 kg
Maximum gross weight:	4 321 kg
Fuel capacity for I hour flying time based on following profile:	
Takeoff at	130% power
Climb	100% power
5 min air combat	130% power
55 min flying time	85% power
Armament:	one MG 151/20 (200 rounds)
	one MG 151/15 (400 rounds)
Equipment:	FuG 18 with D/F
	UV cockpit lighting
	pressurized cockpit

The V1 was completed before the BMW turbojet engines became available, and on 18 April 1941 the aircraft made its first flight powered by a Junkers Jumo 210 G piston engine (730 hp) with Fritz Wendel at the controls. This engine was not sufficiently powerful for high-speed trials, however.

Not until the beginning of 1942 did the BMW turbojets become available. On 25 March 1942 Fritz Wendel took off in the V1, which was now fitted with two BMW 3302 turbojet engines and one Jumo 210 G piston engine. Shortly after takeoff, at a height of 50 meters, both jet engines stopped as a result of turbine blade failure. The Jumo 210 G made it possible for the pilot to complete a circuit and land.

Meanwhile, development of the Jumo 004 engine made it possible for Junkers to deliver the V9 and V10 prototypes to Messerschmitt in July 1942. These were installed in the Me 262 V3, and on 18 July 1942 Fritz Wendel took off on the first pure jet flight by the type. The V1 had made its initial flight from Augsburg, however, Leipheim had been chosen for this flight as it possessed a significantly longer concrete runway.

Preceding taxiing trials had revealed that the Me 262's elevator was ineffective, a result of its tailwheel undercarriage. It was obvious that a tricycle undercarriage was required. In the interim the following takeoff procedure was adopted:

1. Acceleration of the aircraft until a speed of 180 km/h was reached after a roll of 800 meters.

2. Brief application of the brakes to raise the tail off the ground.

This procedure placed the elevator in the air flow, allowing its aerodynamic effects to develop.

It was also found that at high angles of attack (especially in turns) the airflow very quickly broke away from the inner wing. It was decided to increase the chord of the inner wing while adding leading edge slots to permit higher angles of attack and reduce turning radius.

Rechlin test pilot Beauvais was scheduled to test fly the V3 on 11 August 1942. Because of the high air temperature on that day the engines failed to develop full power, and the aircraft was unable to achieve the necessary speed for takeoff. Beauvais succeeded in getting the machine airborne, however, he

Me 262 A-1a jet fighter bearing the code "White 4."

was unable to gain altitude. The machine skimmed over a cornfield and then crashed. This was the first of many more or less serious crashes involving the Me 262. At that time the effects of air temperature on the performance of turbojet engines were not yet known.

At a joint conference of the RLM, the *E-Stelle Rechlin* and Messerschmitt held on 12 August 1942 it was agreed that five more prototypes and ten pre-production aircraft would be built, to be equipped as follows:

- tricycle undercarriage
- pressurized cockpit
- FuG 16Z and FuG 25A
- original armament of three MG 151 to be replaced by one MK 103 and two MG 151, installation of two MG 151 in wings to be examined
- armor for pilot and fuel tanks
- speed brakes

Two further Jumo 004 A-0 power plants were delivered to Messerschmitt in September 1942 and these were installed in the V2. It made its first flight at Lechfeld on 1 October 1942 in the hands of Fritz Wendel. This site was chosen for safety reasons as it offered a longer concrete runway than Leipheim. On the same day Beauvais test-flew the V2 at the *Erprobungsstelle Rechlin*.

With the Me 262 program encountering delays, in January 1943 alternate solutions were sought. Messerschmitt proposed a jet-powered version of the Bf 109. The Bf 109 TL was configured as follows:

- fuselage of the Me 155 (carrier version of the Bf 109)
- wing of the Me 409 (predecessor of the Me 155)
- undercarriage of the Me 309
- two Jumo 004 B turbojets beneath the wings
- revised nose housing two MG 151/20 and two MK 103 cannon
- redesigned tail

It soon became apparent, however, that too many modifications would be necessary, and consequently develop-ment time would not be significantly less than for the Me 262. Therefore, after just two months the Bf 109 TL project was halted.

At the beginning of March 1943 the V1 was refitted with two Junkers Jumo 004 A-0 engines and the Jumo 210 G piston engine was deleted. The aircraft made its first flight in this configuration on 19 July. Repairs to the V3 were completed at the end of March, raising the number of prototypes available for flight testing to three.

On 25 March 1943 Messerschmitt produced a project description incorporating the results of testing to date. For the first time mention was made of employing the Me 262 in the fighter-bomber role, as well as its design role of pursuit fighter. As well, a step-by-step plan for improving the pre-production series was issued.

The *Luftwaffe* general staff first began showing interest in the Me 262 in April 1943. On 17 April *Hauptmann* Späte of *Erprobungskommando 16* (Test Detachment 16) test-flew the Me 262 V2 and was impressed by its performance. This is not surprising, given the fact that this combat pilot had previously flown only propeller-driven aircraft. Influenced by Späte's report, on 22 May 1943 *General der Jagdflieger* Galland flew the V4 at Augsburg. He, too, was impressed by its performance, and immediately became an advocate of the Me 262, pushing for its production and use in the fighter role.

The obvious need for a tricycle undercarriage for the Me 262 was realized in the V5. The prototype was fitted with the nosewheel of the Me 309 in a fixed trials installation. It completed its first flight in this configuration on 6 June 1943.

The Jumo 004 A-0 series engine had been designed using so-called "scarce materials," which caused a number of problems. It was therefore redesigned, resulting in the "scarce materials free" B-series. The first pre-production Jumo

Step-by-Step Plan for Improving the Pre-Production Series
Step 1: State of the aircraft at the beginning of the pre-production series

Purpose:	Pursuit fighter
Armament:	Three MG 151/20 (300 rounds per gun) in fuselage nose
Fuel:	two 900-liter tanks
Undercarriage:	Mainwheels 770 x 370 mm
	Nosewheel 660 x 160 mm
Takeoff weight:	Max. 5 500 kg

Takeoff assist, pressurized cockpit, ejection seat and speed brake deleted.

Step 2:

Purpose:	Fighter-bomber
Armament:	Initially three MG 151/20 in the fuselage nose and two MK 108 in or under the wings, later four MK 108 or two MK 103 and one MG 151/20 in nose compartment
Gravity weapons:	One SC 500 or two SC 250 or one BT 700 torpedo (requires removal of two nose weapons)
Fuel:	Two 900 l (protected) and one 300 l (unprotected)
Undercarriage:	Mainwheels 840 x 300 mm
	Nosewheel 660 x 160 mm
Takeoff weight:	Normal ca. 6 000 kg
	Maximum ca. 7 100 kg

Leading edge slats on entire length of wing, with additional takeoff assist; no pressurized cockpit, ejection seat or speed brake.

004 B-0 turbojets became available in the summer of 1943. First installation of the new engines was in the Me 262 V6. This aircraft was also the first with a retractable tricycle undercarriage, and it made its first flight on 17 October 1943. Meanwhile, on 28 May 1943 the RLM, together with the test detachment, had finalized the construction standard for the first 100 production aircraft, which were designed exclusively for the fighter role.

Experience had shown that the early turbojet-powered aircraft required a relatively long takeoff run. The answer to this was takeoff-assist rockets, which each produced 500 kg of thrust. These were first tested on the V5 on 13 July 1943.

At the end of 1943 the first Junkers Jumo 004 B-1 engines became available. They were installed in the V7, which flew for the first time on 20 December 1943.

In January 1944 Junkers received permission to begin quantity production of the Jumo 004 B-1 engine. Messerschmitt was to simultaneously begin

Construction Standard of the First 100 Production Aircraft

Power plants:	Jumo 004 A, later Jumo 004 B
Fuel:	Two 600 l (protected)
Armament:	Three MG 151/20 (each with 320 rounds). Parallel to this design of an interchangeable nose with six MK 108 or four MK 108 with installation of the Jumo 004 B
Armor:	Against fire from behind, also head and upper arm protection
Undercarriage:	Initially 770 x 270 mm mainwheels and 650 x 150 mm nosewheel, later larger 830 x 300 mm mainwheels
Takeoff assist:	Powder rockets by Rheinmetall-Borsig
Pressurized cockpit:	Pressure-proof components and conduits, but no pressure maintenance equipment
Structural Strength:	H5 with gross weight of 5 000 kg after reaching altitude of 6 000 m. Maximum allowable speed at ground level 900 km/h
Handling qualities:	Elevator forces reduced by 40%, aileron forces by 50% compared to the Me 262 V3.

Speed brake and ejection seat deleted.

construction of the Zero, or pre-production series, at Leipheim, however, this was delayed until March 1944.

The V9 made its first flight on 19 January 1944. Until September 1944 it was used mainly for experiments in the electro-acoustical location of enemy aircraft, and not until October was it converted into the HG I, the first high-speed test-bed based on the Me 262.

Messerschmitt began receiving production Jumo 004 B-1 engines in February 1944. The first engines of this type were installed in the V8, which completed its maiden flight on 18 March 1944. From that point on the V8 was used for trials with an armament of four nose-mounted MK 108 cannon until, on 19 April 1944, it became the first jet fighter to be assigned to *Erprobungskommando Thierfelder* for service trials.

The last of ten prototypes from the original order, the V10 first flew on 15 April 1944. It was the first prototype to be equipped as a fighter-bomber.

The first pre-production aircraft (S-series) were completed by Messerschmitt at Leipheim at the end of April 1944. All but five aircraft were taken into the flight test program to make up for the inevitable losses and thus avoid delays in the program.

The Me 262 in Operational Service

At the beginning of May 1944 the S3 and S4 pre-production machines joined the service trials unit, allowing it to establish a training program. In July 1944 *Erprobungskommando Thierfelder* flew its first missions against Mosquito reconnaissance aircraft. On 26 July 1944 *Leutnant* Alfred Schreiber scored the first victory by a jet fighter in the history of air warfare, shooting down a Mosquito photo-reconnaissance aircraft while flying an Me 262 A-1a (W.Nr. 130 017). After *Hauptmann* Thierfelder was killed in the crash of an Me 262 on 18 July

1944, his test unit formed the core of *Kommando Nowotny*, which in turn led to JG 7, the first service unit equipped with the type.

The *Erprobungsstelle Rechlin* received its first Me 262 on 10 June 1944 and began its own test program. One month later Rechlin had thirteen Me 262s, and by 20 September 1944 the test station had logged 350 flying hours in the course of 800 flights in the Me 262. Based on the results of these tests, at the end of September the *E-Stelle Rechlin* gave the Me 262 A-1a conditional approval for use as a fighter aircraft.

Meanwhile, *Feldwebel* Herlitzius, who had been seconded to the company test program, reached a speed of 1,004 km/h in a steep full-power dive from an altitude of 7,000 meters while flying pre-production aircraft S2 (W.Nr. 130 007, VI + AG).

In the summer of 1944 the Junkers Jumo 004 program again ran into problems, and it was decided to test the redesigned BMW 003 A engine. *Werknummer* 170078 was equipped with BMW 003 A engines, and 130184 and 130188 with BMW 003 A-2 power plants (designation Me 262 A-1b). Trials continued until the beginning of 1945, however, the Me 262 A-1b never entered production, as engine problems persisted and maximum speed never exceeded 800 km/h. Furthermore, the engine situation at Junkers had become less troubled and the BMW power plants were allocated to the Arado Ar 234 C and He 162 programs.

Testing of the V10 for the fighter-bomber role began immediately after its first flight. The aircraft first flew with one 250-kg bomb on 27 May 1944 and one 500-kg bomb in June-July 1944. The Me 262 A-2a differed from the day fighter version (A-1a) in several respects. Two ETC 503 bomb racks, or "Viking Ships," were installed under the fuselage, fixed armament was reduced to two MK 108 cannon, and an additional fuel tank was installed in the fuselage, resulting in a weight increase of 325 kg. The V10 also tested the TSA-2A low-level and dive-bombing system made by Zeiss, which included a partially-programmable bombing computer.

Messerschmitt constructed a total of 28 Me 262 A-2a fighter-bombers in June 1944. A conversion program was established at Lechfeld to retrain bomber pilots to fly the jet fighter-bomber on *"Blitzbomber"* missions with *Erprobungskommando Schenk.*

Production of the fighter-bomber only ran from June to October 1944, when all production switched to fighter aircraft following the introduction of the Emergency Fighter Program.

Operational Fighter Variants

Because of the powerful defensive capabilities of heavy bombers flying in formation and the resulting desire to provide German pilots with an effective means of attacking the bombers while minimizing exposure to defensive fire, the armament of the jet fighter was the constant subject of discussion, development, and testing.

Cannon

The armament of the "pursuit fighter" was originally supposed to be three MG 151/15 cannon. The Me 262 A-1 production version was equipped with an "interim" armament of four MK 108 cannon and the standard Revi 16b gunsight. Plans included a mixed heavy armament of two MG 151/20, two MK 103, and two MK 108 cannon. This tendency toward heavier armament was typified by the Me 262 A-1a/U5. This variant, of which just one example was built (W.Nr. 111 355), was equipped with six MK 108 cannon in the nose weapons compartment. In addition to experiments with

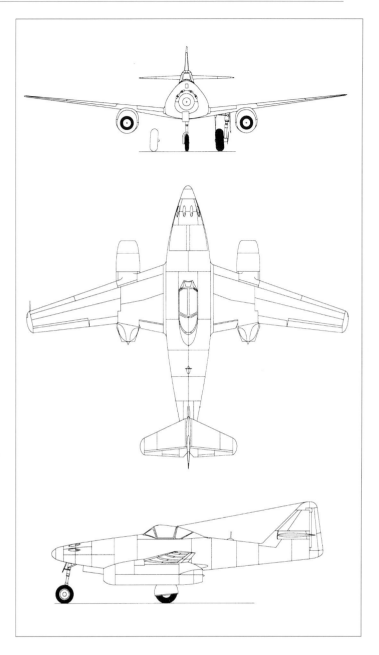

batteries of cannon, an attempt was also made to increase the aircraft's firepower by installing a single large-caliber, long-range weapon in the Me 262. After the BK 5 cannon failed to live up to expectations, at the end of February 1945 W.Nr. 111 899 and 170 083 were each fitted with a single Mauser MK 214 A (50-mm). This Me 262 A-1a/U4 *"Pulkzerstörer"* (formation destroyer) was tested at Lechfeld in March-April 1945.

Messerschmitt Me 262 A-1a

Me 262 A-1a fighter waits to take off, Lager Lechfeld, spring 1945.

Me 262 A-1a fighter (Werk.Nr. 110 604).

Unguided Air-to-Air Rockets

Initial trials with the W.Gr. 21 rocket mortar and the Me 262 did not live up to expectations. Equally disappointing were experiments with the R 100 BS heavy rocket projectile mounted on Me 262 *Werknummer* 111 994 (BS = incendiary shrapnel).

Not until the advent of the R4M rocket was a successful solution found. First use of his armament was on 18 March 1945. Sixty Me 262s were subsequently fitted with racks to carry twelve rockets (each weighing 4 kg) beneath each wing, and six with racks for twenty-four rock-ets beneath each wing. The combination of four MK 108 cannon and R4M rockets endowed the Me 262 with enormous firepower.

Guided Air-to-Air Rockets

In the first months of 1945 *Werknummer* 111 994 was used for trials with the wire-guided Ruhrstahl X-4 air-to-air missile, which was powered by a BMW 109-548 liquid-fuel rocket engine. Test sorties were flown with one missile under each wing. Development was halted in favor of the R4M after the factory producing the liquid-fuel rocket motors was destroyed in an air attack.

Me 262 C-1a V186 "He-imatschützer 1" during takeoff.

Interceptor Fighter

In 1943 a proposal was made for a rocket-boosted version of the Me 262 which could serve as a fast-climbing interceptor. Such a fighter would be capable of intercepting a surprise attack, make several attacks, and pursue the enemy if necessary.

Three versions of the *Heimatschützer* (home defender) Me 262 were proposed at the end of 1943:

- the *Heimatschützer I* with two Jumo 004 C engines and one Walter HWK R II-211 rocket motor, armament of six MK 108 cannon.
- The *Heimatschützer II* with two BMW 003 R engines with BMW 109-718 rocket boosters, armament of six MK 108 cannon.
- The *Heimatschützer III* with two HWK R II-211 rocket motors in place of jet engines, armament of six MK 108 cannon.

Heimatschützer I

Conversion of *Werknummer* 130 186 (V186) into the Me 262 C-1a began on 2 September 1944. A Walter HWK 509 A-2 rocket motor was installed in the rear fuselage, requiring extensive modification of the fuel tank system to accommodate the rocket fuel (*T-* and *C-Stoff*). The high temperature exhaust jet also made it necessary to shorten the aircraft's rudder.

The first ground run of the Walter rocket motor took place on 25 October 1944. On 27 January 1945 Gerd Lindner made the first flight with the rocket motor engaged, reaching an altitude of 8,000 meters in three minutes. Six more flights were made before the aircraft was heavily damaged in a bombing raid on 23 March 1945. After the war the machine was taken to Farnborough where it was examined.

Heimatschützer II

Werknummer 170 074 (V074) was converted into the Me 262 C-2b. The aircraft was powered by two BMW 003 R power plants. The BMW 003 R consisted of a BMW 003 A-1 turbojet engine (800 kg thrust) combined with a BMW 109-718 liquid-fuel rocket motor (1,250 kg thrust).

As a result of problems with sealing and the corrosive rocket fuel, the first satisfactory ground run of the rocket motors did not take place until 23 March 1945. On 26 March Karl Baur made the first flight with the rocket motors engaged, reaching an altitude of 8,200 meters in 1.5 minutes. After this the aircraft made just one more test flight.

Me 262 A-2a fighter-bomber
(Werk.Nr. 110 813) armed with
two SC 250 bombs.

Fighter-Bomber

Two test programs were carried out at the end of 1944 for the purpose of employing the Me 262 in the fighter-bomber role.

The *E-Stelle Rechlin* used *Werknummer* 130 164, 130 188, and 170 070, mainly for dive-bombing trials. The TSA-2A low-level and dive bombing system was also tested, and it proved to be clearly superior to the standard reflector gunsight. Fighter-bomber versions so equipped received the designation Me 262 A-2a/U1.

Messerschmitt used *Werknummer* 170 303 (V303) for bombing trials with external loads as large as a single 1 000-kg bomb.

High-Speed Bomber

At the end of October Messerschmitt began flight testing of *Werknummer* 110 484 (V484). In place of the weapons compartment, this aircraft was fitted with a glazed nose housing a bombardier in a prone position. It was also equipped with a Lotfe 7H sight for horizontal bombing. At the beginning of January 1945 the similarly equipped *Werknummer* 110 555 (V555) joined the test program. Soon afterwards both aircraft were equipped with the K22 fighter autopilot developed by Siemens.

Both aircraft received the designation Me 262 A-2a/U2.

Reconnaissance Aircraft

To make up for the absence of the requested turbojet-powered reconnaissance aircraft, in the summer of 1944 Messerschmitt began converting a small number of Me 262 A-1a fighters into Me 262 A-1a/U3 interim reconnaissance aircraft. The fighter's four MK 108 cannon were replaced with two Rb 50/30 cameras. Once in service, however, the units retrofitted several of these interim reconnaissance aircraft with two MK 108s. In October 1944 an operational detachment dubbed *Einsatzkommando Braunegg* was formed in southern Germany for the tactical reconnaissance

Me 262 A-2a fighter-
bomber.

role, equipped with the Me 262 A-1a/U3. The detachment was later incorporated into NAG 6, a tactical reconnaissance *Gruppe.*

Night Fighter
The good results achieved by the Me 262 against Mosquito reconnaissance aircraft by day made it inevitable that the type would also be used against Mosquito pathfinders at night.

On 18 July 1944 a discussion was held on the installation of various electronic aids in the Me 262. In September 1944 it was decided to produce the Me 262 B-1a/U1 night fighter based on the Me 262 B-1a two-seat trainer. The principal changes would be the installation of electronic equipment and a revised fuel tank arrangement.

In a parallel development, in October-November 1944 the *E-Stelle Rechlin* conducted experiments with a single-seat Me 262 A-1a (W.Nr. 170 095) in the *"Wilde Sau"* night fighter role. On 12 December 1944 Lt. Welter of *Kommando Welter* scored the first victory by a jet night fighter while piloting a slightly-modified Me 262 A-1a (ultra-violet cockpit lighting, map lighting, and a back-up artificial horizon). The first interim night fighters were delivered in March 1945.

Against this background of operational testing, Messerschmitt continued development of the Me 262 B-2a night fighter. The V056 was fitted with radar antennas, and on 9 March 1945 it began flight trials to determine their aerodynamic effects.

Me 262 B two-seat trainer on
display at Willow Grove,
1958.

List of Variants

A-1a	single-seat fighter with four MK 108
A-1a/U1	fighter with two MK 103 and two MG 151/20 or two MK 108
A-1a/U2	bad weather fighter with four MK 108, Siemens K22 autopilot, FuG 120, FuG 125 and EZ 42
A-1a/U3	interim reconnaissance aircraft with two Rb 50/30 cameras, no armament
A-1a/U4	fighter with BK 214 (50-mm)
A-1a/U5	fighter with six MK 108
A-2a	fighter-bomber/high-speed bomber with two MK 108, "Viking Ship" or ETC 503 bomb racks, later ETC 504 for two 250-kg or one 500-kg bomb
A-2a/U1	fighter-bomber/high-speed bomber with two MK 108 cannon and TSA-2A bombsight in the fuselage nose, otherwise as A-2
A-2a/U2	very high-speed bomber with second man in prone position in glazed nose with Lotfe 7H bombsight
A-3a	heavily armored fighter, otherwise as A-1a
A-5a	planned production reconnaissance aircraft with two Rb 50/30 cameras and two MK 108
B-1a	two-seat training aircraft with dual controls, four MK 108
B-1a/U1	interim night fighter with trainer fuselage, four MK 108 cannon and night fighter equipment
C-1a	Heimatschützer I (sometimes also called Me 262 D-1), as A-1a but with additional Walter 109-509 rocket motor in rear fuselage
C-2b	Heimatschützer II, as A-1a but with two BMW 003R engines (BMW 003 A-1 turbojet with additional BMW 109-718 rocket motor)

The suffix a indicates a variant equipped with Jumo 004 engines
The suffix b indicates a variant equipped with BMW 003 engines

Me 262 Development Summary

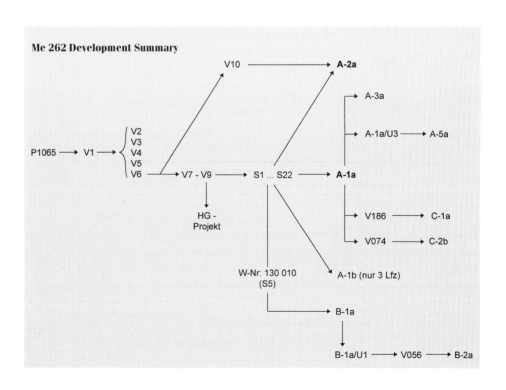

Me 262 B-1a/U1 interim night fighter in display in Hyde Park on Battle of Britain Day in September 1945.

Me 262 A-1a V056 with Lichtenstein AI radar for the night fighter role.

High-Speed Project

As early as 1941 Messerschmitt formulated plans to equip the Me 262 with a 35° swept wing. It was not until July 1943, however, that the idea was taken up in an effort to reduce the effects of compressibility which began appearing at speeds of Mach .75. On 16 February 1944 the Oberammergau project bureau produced a "high speed" design proposal for a very high speed aircraft based on the Me 262. The project was to proceed in three steps:

Me 262 HG 1

Conversion work on the Me 262 V9 (W.Nr. 130 004) began on 1 October 1944. It was fitted with a shallower, streamlined fairing aft of the canopy, swept horizontal tailplanes, and a larger vertical tailplane. Flight testing of the HG 1 began on 18 January 1945 at Lechfeld. The new tail surfaces caused stability problems, and after five flights the standard tail was reinstalled. Twenty more flights were carried out to investigate pressure distribution on the new canopy fairing.

Unpainted Me 262 A-1a
fighter. The various panels
in the airframe structure are
clearly visible (above).

Me 262 A-2a fighter-bomber
(right).

Me 262 B-1a/U1 interim
night fighter (Werk.Nr. 110
306) at Wright AFB in
October 1945. The aircraft
bears the foreign equipment
code FE 110 (below and
facing page).

Me 262 A-2a at Wright AFB in October 1945. The aircraft's bomb racks have been removed (above).

Restored Me 262 B-1a/U1 interim night fighter (Werk.Nr. 110 305) on display in the National War Museum in Johannesburg, South Africa (right).

Me 262 A-1a V167 prior to takeoff (facing page).

Me 262 A-1a V167 (Werk.Nr. 110 167). It replaced the V5 after it was destroyed in a crash and went on to complete 300 flights, including stability test flights and trials with the EZ 42 gunsight.

Me 262 HG II

The next stage involved the conversion of Me 262 *Werknummer* 111 538. In addition to the measures adopted for the HG I, the wing was swept at 35° on the quarter-chord line.

The aircraft was ready to fly, however, it was seriously damaged in a ground accident and was not repaired prior to the end of the war.

Messerschmitt HG III

The only feature shared by this design and the Me 262 was the latter's fuselage. Design changes to the HG III included:

- 45° swept wing (46.5° at the quarter-chord line) using the outer wing panels of the production Me 262.
- power plants (two Junkers Jumo 004 D or two Heinkel HeS 011) buried in the wing roots.
- main undercarriage attached at the wing roots retracting forward to lie beneath the cockpit in the fuselage center section.
- swept tail as on the HG I, later swept vee-tail.
- special cockpit for high speed trials.

The aircraft was designed to achieve speeds of 1,050 km/h at ground level and 1,100 km/h at 6,000 meters. When the war ended the project was still in the development stage, and extensive wind tunnel investigations had been carried out.

Prototypes

The V1 to V10 prototypes were constructed by Messerschmitt at Augsburg-Haunstetten.

The pre-production series (S-series) was built at Leipheim. Of the twenty-two examples built, only the S1 to S5 were used in their intended role, while the rest joined the flight test program as replacements for crashed prototypes or were assigned prototype (V) numbers.

Following the loss of the V1, V2, V4, V5, V6, V7, and V8 prototypes during

List of Prototypes

WNr.	Code		First Flight	Remarks
V1	Me 262 00 001	PC+UA	18/04/1941	with one Jumo 210 G
			25/03/1942	with one Jumo 210 G and two BMW 3302 Twk
			19/07/1943	with two Jumo 004 A and three MG 151/20
V2	Me 262 00 002	PC+UB	01/10/1942	with two Jumo 004 A-0
V3	Me 262 00 003	PC+UC	18/07/1942	first pure jet flight by an Me 262 with two Jumo 004 A-0 engines
V4	Me 262 00 004	PC+UD	15/05/1943	with two Jumo 004 A-0
V5	Me 262 00 005	PC+UE	06/06/1943	with two Jumo 004 B-0, fixed nosewheel from Me 309
V6	130 001	VI+AA	17/10/1943	first hydraulically-retractable nosewheel, two Jumo 004 B-0 engines
V7	130 002	VI+AB	20/12/1943	two Jumo 004 B-1 engines, fittings for pressurized cockpit
V8	130 003	VI+AC	18/03/1944	two Jumo 004 B-1 engines and four MK 108, used for weapons trials
V9	130 004	VI+AD	19/01/1944	production prototype for the A-1a fighter: no armament, from Oct. 1944 modified for use in high-speed project
V10	130 005	VI+AE	15/04/1944	production prototype for the A-2a fighter-bomber series
S1	130 006	VI+AF	19/04/1944	weapons trials
S2	130 007	VI+AG	28/03/1944	high-speed experiments (achieved 1 004 km/h on 25/06/44)
S3	130 008	VI+AH	April 1944	nosewheel development and testing
S4	130 009	VI+AI	May 1944	horizontal tail development, equipped with Flettner tabs
S5	130 010	VI+AJ	Apr. 1944	two-seater prototype

Experimental Aircraft

	WNr.	First Flight	Remarks
262/V1	130 015	30/06/1944	S10 of the pre-production series, used for various tests including a raked windscreen (25° instead of the usual 33°)
262/V2	170/056	June/July 44	investigations of directional stability, later fitted with FuG 218 and FuG 226 radar sets
262/V4	170 083	autumn 1944	initially used for undercarriage tests, later modified as "Pulkzerstörer" with BK 214 cannon, March 1945
262/V5	130 167	31/05/1944	fighter-bomber trials, equipped with EZ 42
262/V6	130 186	16/10/1944	prototype for C-1a series: first flight with rocket booster on 15/02/1945
262/V7	170 303	autumn 1944	fighter-bomber trials with bomb loads up to 1 000 kg
262/V8	110 484	Oct. 1944	prototype for A-2/U2 with glazed nose
262/V11	110 555	winter 44-45	prototype for A-2/U2 with glazed nose
262/V12	170 074	08/01/1945	two BMW 003 R turbojets; prototype for

C-2b series; first flight with rocket boosters on 26/03/1945

testing, in November 1944 it was decided to replace them with new aircraft.

The new prototypes were initially assigned the V-number of the aircraft they replaced (for example, V2 or V5), however, later prototypes were assigned designations consisting of a V followed by the last three digits of the aircraft's *Werknummer* (for example V056 or V167).

Production

In a report prepared for the Americans in June 1945, Messerschmitt stated that a total of 1,433 Me 262s were built by 19 April 1945. The following table provides a breakdown of production by the Messerschmitt AG Augsburg (A), with the Leipheim, Kuno, and Schwäbisch Hall assembly facilities, and the Messerschmitt GmbH Regensburg (R), with the Obertraubling and Neuburg/Donau assembly facilities.

Deliveries of Me 262 Aircraft

Month	1944 A	R	1945 A	R
January			163	65
February			166	130
March	1		165	75
April	15		64	37
May	7			
June	28			
July	58			
August	15			
September	92	2		
October	108	10		
November	87	14		
December	108	23		
Totals:	519	49	558	307
Grand total:		1,433		

Of these 1,433 aircraft, 240 were produced as bombers from April to October 1944

Messerschmitt Me 328

At the beginning of 1941 Messerschmitt began preliminary design work on a fast single-seat, twin-engined light bomber which could also be used as a fighter, shipborne fighter, or reconnaissance aircraft. The aircraft was designed around the Schmidt-Argus As 014 pulse-jet, which was still under development. With the air battle over England still raging, Messerschmitt was given a contract to construct three prototypes, construction of which began in September 1941. For capacity reasons, at the beginning of 1942 production was switched to DFS at Ainring, which produced the welded steel tube fuselage and attached the wooden wing and the tail assembly (taken from the Bf 109 G).

After initial unpowered trials in the summer of 1942 at Ainring, the first two prototypes were transferred to Hörsching airfield near Linz, where the As 014 pulse jets were installed in different configurations (under the wings, on the fuselage). Flight trials revealed several problems with the pulse jets:

- they developed just 150 kg of thrust, and

- performance depended not just on height, but on speed, as well.

In contrast to investigations by Argus, it was discovered that performance of the pulse jets fell off as speed increased.

Nevertheless, in autumn 1942 the Me 328 was included in the RLM's call for tenders for a high-speed bomber. As well, the RLM (GL/C-E2) issued technical specifications for the Me 328 B high-speed bomber (a derivative of the Me 328 A shipborne fighter) (see box below).

Messerschmitt developed a production program for 300 Me 328 B aircraft dated 15 December 1942. The aircraft was designed as "a cantilever mid-wing monoplane of wooden construction with two Argus jet power plants, each developing 300 kg of thrust, enclosed cockpit, landing skids, and external bomb load." Twenty aircraft were to be built as prototypes and 280 as production aircraft by the Jacobs-Schweyer Flugzeugbau GmbH in Darmstadt. Projected delivery date was June 1943.

Further problems with the Argus pulse jets caused testing to drag on well

Me 328 High-speed Bomber Technical Requirements

Crew:	One
Purpose:	Coastal defense within specified radius of action. Because of power plant characteristics missions limited to low-level (below 10 000 m) attacks against enemy shipping and landing targets (southern England), attack horizontally or in an approximately 20-degree dive (toss bombing).
Power plant:	two Argus pulse-jets, initially 300 kg of static thrust each, later 400 kg. Easiest possible engine change because of short power plant operating life.
Performance:	With two 300-kg Argus pulse-jets and one 500 kg bomb at ground level: 600 km/h
	With two 400-kg Argus pulse-jets and one 1 000 kg bomb at ground level: 700 km/h
	Maximum landing speed 180 km/h on skid. Special braking for wet fields and snow.
	Endurance one hour
	Range 600 km
Takeoff:	It is expected that the same takeoff trolley will be used as developed for the Me 163 B, meaning takeoff using the takeoff trolley with jet propulsion on a standard gauge track. Medium acceleration 2g. Takeoff run on a track, at least one minute. Otherwise towed takeoff (using Me 163 B takeoff trolley).
Construction:	The airframe is to be constructed mainly of wood.
Design load:	Stress Group H 5. Safe load coefficient of 5 for target weight with bomb.

into 1943. It was found, for example, that sonic pressure from the very noisy pulse-jets caused damage to the aft fuselage. For this and other reasons the project was canceled on 3 September 1943.

The Me 328 was revived in early 1944. An Allied invasion on the Channel Coast was expected, and some *Luftwaffe* pilots wanted to employ a suicide weapon, specifically a manned glider-bomb, against it. The pilot was supposed to steer the aircraft toward a ship and, if possible, bail out at the last minute. The Me 328 was considered, because it was believed that it could be made available for this role in a short time. Instead of a bomb, a 500-kg torpedo warhead would be fitted in the aircraft's nose.

Testing of the Me 328 for the suicide role continued until April 1944, when quantity production was supposed to begin. Among those who participated in the trials was Hanna Reitsch, who wrote:

"The Me 328 was originally conceived as a fighter or destroyer. Now we were to use it as a kind of manned glider-bomb on suicide missions. It was a single-seater with stub wings, about four to five meters in span. Glide ratio was about 1 : 12 at 250 km/h, and about 1 : 5 at 750 km/h. The Me 328 could not take off on its own, instead during trials it was carried piggyback fashion on the wing of a Do 217 to an altitude of 3,000 to 6,000 meters. From the cockpit it was possible to release the Me 328 from the Do 217, and it was a simple matter to raise it from the wing in flight. Handling qualities were sufficient for its planned role. We had to demand a good view, comfort, good maneuverability, longitudinal stability, and directional stability. These conditions were met."

In June 1944, however, the RLM decided on another suicide aircraft, a manned version of the Fi 103 flying bomb. The Fi 103 had been under test since December 1942 and was powered

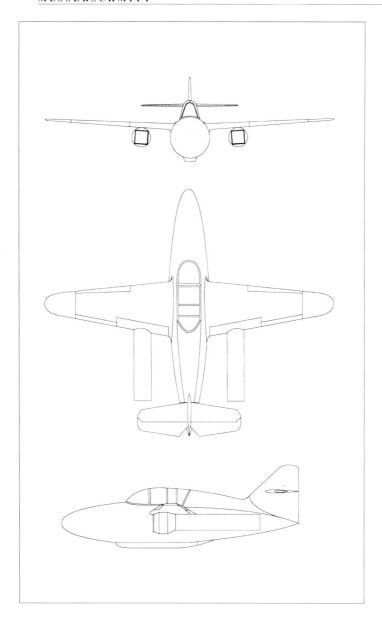

by a single As 014 pulse-jet. The V1 and V2 were the only examples of the Me 328 to be completed. When the program was canceled the V3 to V10 prototypes were still under construction. The Me 328's only claim to fame was that it was the first aircraft to be powered by pulse jets.

Messerschmitt Me 328

Me 328 with Argus As 014 pulse-jets.

Messerschmitt P1101

In mid-1944 the *Luftwaffe* High Command issued a development contract for a new fighter aircraft. It was to be powered by a single Heinkel HeS 011 turbojet engine and achieve a speed of 1,000 km/h at an altitude of 7,000 meters. Armament was to consist of four MK 108 cannon with 60 rounds per gun.

The participating firms—Blohm & Voss, Focke-Wulf, Heinkel, and Messerschmitt—were brought together by the RLM on 10 September to establish a common basis on which to calculate performance figures for the turbojet fighter designs. As a result of this meeting, Junkers and Henschel joined the list of participants, and Messerschmitt revised its initial design of the P1101. It was based on an earlier project, the P1092 of 1943, however, it had a wing with 40 degrees of sweepback. On 30

Me P1101 in Oberammergau.

Me P1101 in the hands of
Bell with mockups of six MG
151 cannon.

September 1944 Professor Messerschmitt decided to build an experimental aircraft with which to evaluate the risks and problems associated with swept wings.

Planning of the military version was to continue. An additional requirement was that as many components of the Me 262 as possible were to be used.

The P1101 was laid out as a mid-wing monoplane with sharply swept wings and tail surfaces. The sweep angle of the wings could be set between 35° and 45° (along the quarter-chord line) on the ground. The best results were expected at a sweep angle of 40°. When it became apparent that development of the Heinkel HeS 011 engine was taking longer than expected, it was decided to equip the prototype with a Junkers Jumo 004 B turbojet producing 900 kg of static thrust.

The Messerschmitt P1101
provided the inspiration for
the Bell X-5 with variable-
sweep wings.

The P1101 V1 prototype was 80 percent complete in April 1945 when it was captured by the Americans at Oberammergau, still engineless. On 7 May a team of American experts of the Air Technical Intelligence Team arrived in Oberammergau; they immediately realized that they were looking at the prototype of a high-speed fighter with swept wings. Among the specialists of the Combined Advanced Field team was Robert J. Woods, chief engineer and cofounder of the Bell Aircraft Corporation. The P1101 was transported to America where, after inspection and evaluation at Wright Field in Dayton, Ohio, it was handed over to Bell Aircraft Corp. At first Bell proposed to the U.S. Air Force that the P1101 program be revived, however, this was rejected, as the aircraft was considered too small to accommodate the necessary armament and fuel.

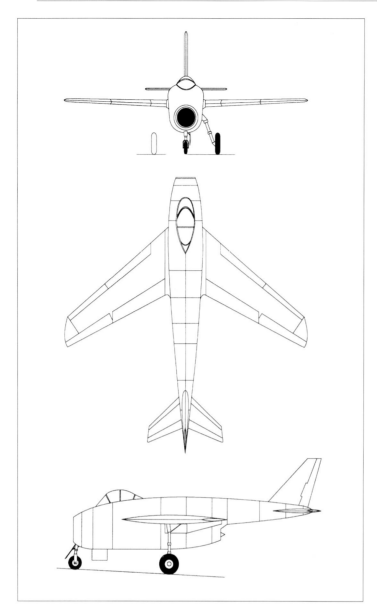

In 1949 Bell proposed the construction of a high-speed experimental aircraft with variable sweep wings based on the P1101. It would be possible to vary wing sweep in flight between 20° and 60°, the process taking approximately 20 seconds. In July of the same year Bell received a contract to develop and build two prototypes to be designated X-5.

The first X-5 flew for the first time on 20 June 1951, the second on 10 December 1951. Interest in swept wings had meanwhile grown, and the X-5s were put through an extensive flight test program. The surviving X-5 (the second machine crashed in 1953) was retired in 1955 and today is on display at the USAF Museum in Dayton.

Messerschmitt Me P1101.

Description of the Messerschmitt P1101

Purpose:	single-seat fighter
Crew:	pilot on ejection seat in pressurized cockpit covered by a bubble canopy.
Wing:	cantilever mid-wing monoplane. Sweep angle adjustable on ground between 35° and 45°. Leading edge slats and landing flaps between ailerons and fuselage. Single-spar wooden construction.
Fuselage:	all-metal monocoque structure, forward section with oval cross-section, aft section circular tail section bearer.
Tail surfaces:	conventional swept cantilever structure. All surfaces swept 45° on the leading edge. All surfaces of all-wood construction.
Undercarriage:	retractable tricycle undercarriage. Mainwheels retract rearward into side of fuselage, nosewheel rearward into fuselage nose with nosewheel turning through 90°.
Power plant:	one Heinkel HeS 011 A turbojet producing 1 300 kg of static thrust. Fuel capacity 935 kg in two wing and one fuselage tanks.
Armament:	two MK 108 in fuselage on either side of air intake.

The Arado High-Speed Bomber
and Reconnaissance Aircraft

Arado Ar 234 *Blitz*

In late 1940 the RLM issued a contract for development of a jet-powered reconnaissance aircraft with a range of 2,200 km to replace the Ju 88 P. This contract was also issued to Arado, which subsequently developed a number of designs which, along with those submitted by the other participating companies, were evaluated by the RLM in early 1941. Arado's E 370 design was accepted, and Arado received a development contract.

Design work began in early 1941. As a result of delays in the development of turbojet engines by Junkers and BMW, initial taxiing trials could not begin until March 1943, the V1 having been fitted with Jumo 004 A-0 engines. Finally, on 15 June 1943 the V1 made its first flight from Rheine airfield. It was piloted by *Flugkapitän* Selle, who later lost his life in the crash of the V7.

It was feared that a retractable undercarriage would not be able to absorb the high forces on landing, therefore, the first eight prototypes were equipped with a central retractable landing skid and two stabilizing skis, one under each engine nacelle. The aircraft took off on a trolley, which was jettisoned after liftoff. This takeoff trolley was the source of numerous problems. In the beginning the trolley was jettisoned too high and the first two were destroyed. The next flights saw the trolley jettisoned from a height of 60 meters. If it was jettisoned too low, the danger existed that the bouncing trolley might strike the aircraft. The trolley was therefore fitted with a braking parachute which deployed when it was jettisoned. From very early on it was apparent that the takeoff trolley was an unsatisfactory solution.

The V2, which was identical to the V1, made its first flight on 27 July 1943. It was followed on 25 August by the V3, which had a pressurized cockpit, an ejection seat, and takeoff-assist rockets. A summary of the V3's performance appears below:

Maximum speed: altitude of 6 000 m	864 km/h at an
Range:	1 380 km
Service ceiling:	12 500 m

The V3 sustained serious damage on its very first flight. On 15 September 1943 the V4, which was identical to the V3, completed its maiden flight. By this time the somewhat less powerful BMW 003 turbojet was also available. The V5 was fitted with two of these power plants, and was able to take off on its first flight on 20 December 1943.

Arado Ar 234 V6 with four
BMW 003 A-1 turbojets.

Arado Ar 234 B-2 in RAF
markings after capture in
Norway.

Another Ar 234 B-2 captured
in Norway seen during pre-
flight.

The prototypes V1 to V5 were very easy to fly. Since no aerodynamic problems were encountered, it was clear that the airframe was capable of even higher speeds until the machine reached its limiting mach number, where the effects of compressibility made themselves felt. It was therefore decided to construct two more prototypes, each of which was to be equipped with four BMW 003 turbojets (each 800 kg static thrust). The V6 had the four engines arranged separately under the wings, while on the V8 they were installed in pairs. The V8 made its first flight on 1 February 1944, becoming the world's first four-engined jet aircraft. The V6 followed on 8 April 1944. The V6 encountered problems in the high-speed range, since the air flow between the airframe and the engine nacelles reached near-supersonic speeds, producing shockwaves. The engine nacelles were moved outboard 130 mm, after which the effect was not felt until a speed of 980 km/h. This speed could only be reached in a dive, however.

Meanwhile, the V7 was completed. It was the A-series prototype. During

one of the first test flights, however, the compressor blades of one of the engines fractured and control cables were severed. The V7 was near the airfield when this happened and subsequently crashed.

Given the problems with the takeoff trolley described earlier and the fact that the aircraft was immobile after landing and could not maneuver on its own while on the takeoff trolley, it was decided to provide the machine with a tricycle undercarriage. This caused a number of problems, however.

The Ar 234 had been designed as a high-wing monoplane with a very thin wing. There was no room in the wing for an undercarriage, and in any case, this would have meant very long undercarriage legs. The designers were forced to widen the Ar 234's slender fuselage slightly to accommodate the main-wheels, but the result was a very narrow-track undercarriage. This and a number of other improvements resulted in the Ar 234 B.

The V9 was the first prototype of the B-series and made its first flight on 12 March 1944. The V9 was equipped with two Junkers Jumo 004 B engines, a pressurized cockpit, an ejection seat, and braking parachute. In March 1944 Walter Kröger achieved a speed of 895 km/h in level flight in the V9. The V10, which was identical to the V9 apart from the pressurized cockpit, followed just three weeks later on 2 April 1944. The following V11 and V12 were similar to the V9 and V10.

Quantity production of the Ar 234 B began a short time later in Alt-Lönnewitz, near the German-Czech border. The first pre-production aircraft were completed and test flown at the beginning of June 1944. The test station at Rechlin received the first thirteen Ar 234 B-0 pre-production machines for extensive testing. With the exception of newly-added military equipment, these pre-production aircraft were similar to the V9 to V12 prototypes.

The B-0 series had no pressurized cockpit, no ejection seat, and was unarmed. Provision was made for the installation of camera equipment in the rear fuselage:

Starting the engine of an Ar 234 B-2 captured in Norway.

- two Rb 50/30 or
- two Rb 75/30 or
- one Rb 70/30 and one Rb 20/30

The B-1 series was similar to the pre-production B-0 and entered service in autumn 1944 in the long-range reconnaissance role.

The V5 and V7 (rebuilt) began flying operational sorties in July 1944. These two aircraft were assigned to *1. Staffel* of the *Versuchsverband des Oberbefehlshabers der Luftwaffe* (Experimental Unit of the *Luftwaffe* Commander in Chief) and were based at Juvincourt near Reims, from where they made reconnaissance flights over England. By August 1944 the V5 had logged 22 hours and the V7 24 hours. On 27 July 1944 the unit was transferred to Chievres, and on 5 September 1944 to Rheine, where a short time later it received two Ar 234 B-1s. By 1 November 1944 the unit had completed 24 reconnaissance flights from Rheine with the Ar 234 B-1, most of them over the east coast of England, as it was feared that the Allies were going to invade Holland. In September 1944 *Kommando Götz* was formed at Rheine with four Ar 234s, including the V6 and V8. Two further reconnaissance detachments were formed at Rheine at the end of November 1944, *Sonderkommando Hecht* with one Ar 234 B-1 and *Sonder-*

kommando Sperling with five Ar 234 B-1s (*Sonderkommando* = Special Detachment). In January 1945 these three units were combined to form 1.(F)/100 under the command of *Luftwaffenkommando West*, which was later joined by 1.(F)/123 and 1.(F)/33 in Denmark. On 10 April 1945 an Ar 234 B-1 from Rheine flew the *Luftwaffe*'s last reconnaissance mission over Great Britain (northern Scotland).

Allied fighters repeatedly tried to intercept and shoot down the Ar 234 reconnaissance machines over Great Britain, however, the Ar 234's superior speed always allowed it to elude its pursuers. Soon after production of the Ar 234 B-1 reconnaissance aircraft began, it was halted in favor of the B-2 high-speed bomber version. A description of the B-2 appears below.

KG 76 was the first *Geschwader* to convert to the Ar 234 B-2. The conversion of Ju 88 bomber pilots of II./KG 76 to the Ar 234 began in the late summer of 1944 at Alt-Lönnewitz. After a few hours in a bomber trainer the pilots made their first flight. Operation of the Jumo 004 power plants was a constant source of problems. There were variations in power output, and the pilots were forced to jockey the throttles to compensate. Unlike piston engines, the

Description of the Production High-Speed Bomber

Purpose:	high-speed bomber, long-range reconnaissance aircraft
Crew:	one
Wing:	cantilever all-metal shoulder-wing monoplane, one-piece two-spar wing, trapezoid shape with very thin cross-section, hydraulically-actuated split flaps, when flaps lowered ailerons automatically droop 10°
Fuselage:	all-metal monocoque structure, ejection seat and braking chute
Tail surfaces:	cantilever centrally-mounted conventional tail surfaces with mechanically-adjustable horizontal stabilizer and rudder trim, control surface mass balances
Undercarriage:	retractable tricycle undercarriage with single-leg struts. Nosewheel retracts rearward into fuselage, mainwheels into fuselage sides.
Power plants:	two Junkers Jumo 004 B-1 turbojets beneath the wings each producing 900 kg of static thrust. Fuel capacity 1 750 + 2 000 liters internally in fuselage and possible use of two 300-l drop tanks
Military equipment:	three 500-kg bombs (one under the fuselage and each engine) or one 1 000-kg bomb (under the fuselage), Lotfe 7K bombsight for horizontal bombing, BZA 1 sight with forward and rearward looking periscopes for dive-bombing. LKS 7D-15 autopilot (necessary for use of Lotfe 7K).

early jet engines had to be handled gingerly. Furthermore, the pilots had difficulties adjusting to the narrow-track undercarriage, which proved difficult in turns.

Various equipment sets (*Rüstsätze*) were developed for the Ar 234 B-2:

b   two reconnaissance cameras

p   Patin three-axis autopilot

r   pick-up points for two 300-l external fuel tanks

All three equipment sets were always installed for the reconnaissance role, and the photo-reconnaissance version was designated the Ar 234 B-2/bpr. The Ar 234 B2/1 was employed as a pathfinder on bombing missions.

The Ar 234 B-2 was part of the RLM's priority program, which also included the Do 335, He 162, and Me 262. A total of 150 aircraft were delivered by the end of 1944. Production of the B-2 series totaled 210 aircraft by the end of the war.

The Ar 234 exhibited excellent flight characteristics, however, handling on the ground was not as good on account of the type's narrow-track undercarriage. Takeoff was very easy; above 60 km/h the aircraft maintained direction without rudder assistance, even in a crosswind. With 160 km/h showing on the airspeed indicator the nosewheel left the ground, and as speed increased the machine lifted off on its own. If the pilot attempted to lift off too soon, the aircraft simply touched down again. Takeoff distance at a weight of 9,350 kg (bomber with three 500-kg bombs) was 1,780 meters. Takeoff assist rockets (*R-Geräte*) were normally used at weights in excess of 9,000 kg, reducing takeoff roll to 690 to 860 meters.

Time to climb to 6,000 meters was 17.5 minutes. When the autopilot was switched on, heading and especially altitude had to conform to the selected height setting, otherwise a sharp "climb" or "descend" command would result which might overstress the airframe. Approach to the target was made

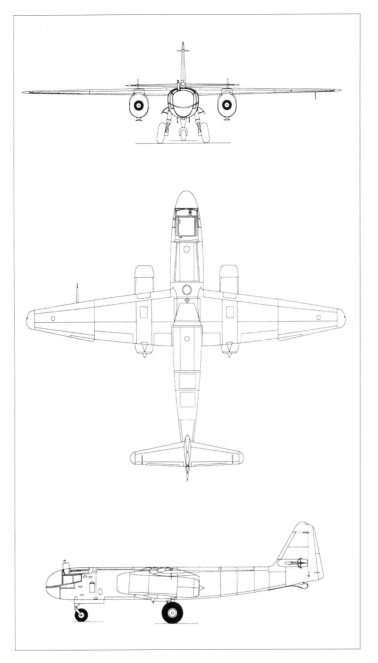

Arado Ar 234 B-2.

at an altitude of 9,000 to 10,000 meters, and speed after bombs were dropped was approximately 700 km/h, 740 km/h at 6,000 meters. The turbojet's third ejector nozzle setting was reserved for emergency power and could only be used for short periods at maximum revolutions above 6,000 meters. The resulting increase in speed was 40 to 50 km/h. Behavior with one engine out was innocuous, and turns into the dead engine posed no problems. Care had to be taken, however, to ensure that fuel was

Arado Ar 234 C-1 with four BMW 003 A turbojets.

Arado Ar 234 C-1.

drawn equally from the two tanks to avoid shifts in the center of gravity. At medium weight with one engine out service ceiling dropped to 5,400 meters and speed to 450 km/h.

Landing with both engines operating was very easy:

• 200 to 210 km/h to the airfield boundary

• throttle levers to idle

• touchdown at 160 to 170 km/h

• unrestricted use of brakes after nosewheel on the ground

Roll-out after landing was longer on grass surfaces than on concrete, as the braking effect was less. The braking parachute was used for landings at smaller airfields. It was possible to go around on one engine with undercarriage and flaps down. In order to protect the pilot, belly landings were to be made with the nosewheel extended if possible. (Technical data from company manual

2234B-2/Fl)

After production of the B-2 had stabilized, Arado began further development work. Like the V8, the V13, which made its first flight on 30 August 1944, was equipped with four BMW 003 A turbojets arranged beneath the wings in pairs. The V14 was equipped with two Jumo 004 engines and represented the B-0 series. It was used to test military equipment. The V16 was supposed to be fitted with two BMW 003 R power plants (BMW 003 A turbojet producing 800 kg static thrust and BMW 109-718 rocket motor). The V16 was intended for high-speed experiments. For this purpose, four different wings were built for the V16:

• one swept wing of wooden construction with decreased sweep in outer panels (crescent-wing as used on the Handley Page Victor)

• one straight wing

- two swept wings with laminar pro
file

Since the BMW 003 R engines were never delivered, the V16 was never tested. Both the V17 (two BMW 003 A-1 engines and pressurized cockpit) and V18 (swept wings and four BMW 003 A-1 engines) were still under construction when the war ended.

The good results achieved with the Ar 234 B-2 and the V8 and V13 prototypes resulted in the development of the C-series. It was to be similar to the B-series, but equipped with four BMW 003 A-1 turbojets. Development work was hampered by effects of the war, and the fuel shortage complicated local transfers of the work.

The V19, the first C-series prototype, made its maiden flight on 30 September 1944. For experimental purposes it was equipped with speed brakes in an attempt to find an alternative to the braking chute. Two variants were projected and developed:

Ar 234 C-1   High-speed reconnaissance aircraft with pressurized cockpit; two cameras in rear fuselage, and defensive armament of two fixed rearward-firing MG 151/20 cannon.

Ar 234 C-2   High-speed bomber; like C-1, but without defensive armament, bomb load of one 1,000 kg bomb or two 500-kg bombs.

The C-1 and C-2 variants were shelved in favor of the multi-purpose C-3. The C-3 was designed as a multi-role combat aircraft. It was supposed to be powered by four BMW 003 C power plants, each producing 900 kg of static thrust, however, these were not available and the BMW 003 A-1 (800 kg of static thrust) was used instead. The cockpit roof was raised to provide a bet-

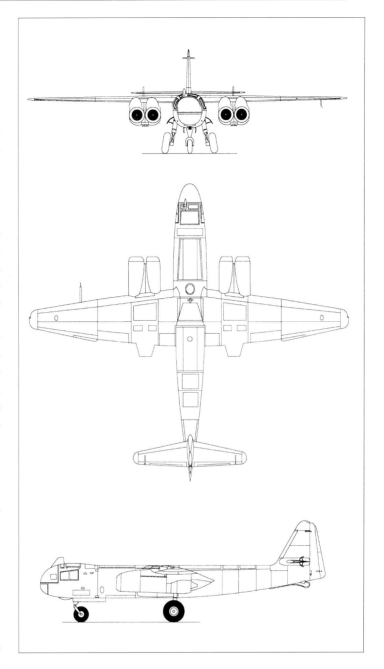

ter view for the pilot, and a fixed defensive armament of two rearward-firing MG 151/20 cannon was planned. Bomb load was similar to that of the C-2. An equipment set would also be available for the night-fighter role, consisting of two MG 151/20 cannon mounted beneath the fuselage.

Nineteen examples of the C-3 series were completed by the end of the war, however, none saw action.

Arado AR 234 C-3.

59

List of Prototypes

	Werk-Nr.	Code	First Flight	Power Plants	Remarks
V1	130 001	TG+KB	15/06/1943	two Jumo 004 A-0	landing skid
V2	130 002	DP+AW	27/07/1943	two Jumo 004 A-0	landing skid
V3	130 003	DP+AX	25/08/1943	two Jumo 004 A-1	1st A-series prototype; ejection seat, pressurized cockpit, takeoff assist rockets, landing skid
V4	130 004	DP+AY	15/09/1943	two Jumo 004 A-1	ejection seat, pressurized cockpit, landing skid
V5	130 005		20/12/1943	two Jumo 004 B-0	landing skid
V6	130 006	GK+IW	08/04/1944	four BMW 003 A	engines housed individually beneath wings; landing skid
V7	130 007	GK+IX	22/06/1944	two Jumo 004 B-1	prototype for A-series; landing skid
V8	130 008	GK+IY	01/02/1944	four BMW 003 A	engines in pairs beneath wings; landing skid
V9	130 009	PH+SP	10/03/1944	two Jumo 004 B-1	1st B-series prototype; ejection seat, braking chute, pressurized cockpit; retractable undercarriage for the first time
V10	130 010	PH+SQ	02/04/1944	two Jumo 004 B-1	ejection seat, braking chute; bombing trials
V11	130 021	PH+SR	05/05/1944	two Jumo 004 B-1	ejection seat, braking chute; engine trials
V12	130 022	PH+SS	15/09/1944	two Jumo 004 B-1	ejection seat, braking chute
V13	130 023	PH+ST	30/08/1944	four BMW 003 A-1	engines paired in nacelles under the wings
V14	130 024	PH+SU	Dec. 1944	two Jumo 004 B-1	equipment Trials
V15	130 025	PH+SV	27/07/1944	two BMW 003 A-1	engine trials
V16	130 026	PH+SW		two BMW 003 R	high-speed flight trials
V17	130 027	PH+SX	25/09/1944	two BMW 003 A-1	engine trials
V18	130 028	PH+SY		four BMW 003 A-1	wing tests
V19	130 029	PI+WX	30/09/1944	four BMW 003 A-1	1st C-series prototype; pressurized cockpit, speed brake
V20	130 030	PI+WY	05/11/1944	four BMW 003 A-1	C-1 series prototype; pressurized cockpit and double glazing
V21	130 061	PI+WZ	16/10/1944	four BMW 003 A-1	C-3 series prototype
V22	130 062	RK+EL	16/12/1944	four BMW 003 A-1	C-3 series prototype
V23	130 063	RK+EM	winter 44-45	four BMW 003 A-1	C-3 series prototype, trials with pressurized cockpit
V24	130 064	RK+EN	12/01/1945	four BMW 003 A-1	C-3 series prototype, trials with pressurized cockpit
V25	130 065	RK+EO		four BMW 003 A-1	C-3 series prototype
V26				four BMW 003 A-1	laminar flow tests in high speed range
V27				four BMW 003 A-1	speed brake Trials
V28				four BMW 003 A-1	C-7 prototype
V29				four BMW 003 A-1	C-7 prototype
V30				four BMW 003 A-1	wing tests
V31				two HeS 011	D-series prototype
V32				two HeS 011	D-series prototype
V33					
to V40					D-series prototypes, not completed

Additional versions of the C-series were in the planning stages:

C-4 reconnaissance aircraft derived from the C-3

C-5 two-seat bomber with pressurized cockpit

C-6 reconnaissance aircraft, otherwise similar to the C-5

C-7 night-fighter, otherwise similar to the C-5

Four additional prototypes, the V26 to V29, were built. The V26 was completed on 20 March 1945 and was supposed to be used for investigation of the laminar flow at very high airspeeds. Building on experience gained with the V19, the V27 was planned as a test-bed for a speed brake. The V28 and V29 were completed before the end of the war and were planned as test-beds for the two-seat Ar 234 C-7 night-fighter.

All of these prototypes were powered by four BMW 003 A-1 turbojets, whereas production aircraft were to be equipped with two Heinkel HeS 011 (1,300 kg static thrust) or two Jumo 004 C (1,200 kg static thrust with afterburning) engines.

There was also a projected C-8 version, a bomber powered by two Jumo 004 D engines (930 kg static thrust).

Development of the D-series ran parallel to later work on the C-series. Two variants were planned and designed, both to be powered by two Heinkel HeS 011 engines:

D-1 two-seat reconnaissance aircraft

D-2 two-seat bomber

Of the ten D-series prototypes planned, two, the V31 and V32, were completed by the end of the war. The Ar 234 P was a night-fighter derived from the D-series with a modified fuselage nose designed to accommodate airborne intercept radar.

The only variants of the Ar 234 to see operational service were the Ar 234 B-1 and Ar 234 B-2. 210 examples were built, but only part of these reached KG 76.

In spite of their small numbers, the Ar 234s of KG 76 were a thorn in the side of the Allies, since they were almost immune to interception. It was not for nothing that the Allies dubbed the Ar 234 the "Blitz" (lightning).

Multi-Engine Jet Bomber with Forward-Swept Wing

Junkers Ju 287

At the end of 1942 the RLM called for the development of a multi-engined jet bomber whose maximum speed was to exceed that of contemporary piston-engined fighters.

At that time the DVL (German Aviation Experimental Institute) had results which indicated that a wing swept at about 35° would make it possible for higher speeds to be achieved by delaying the formation of shock waves in the transonic speed range, meaning that a higher critical Mach number could be achieved with a swept wing than with a straight one.

At the same time, as part of Project EF 116, Professor Heinrich Hertel of Junkers was carrying out experiments with a negatively-swept wing (sweep angle 23.5°), which was expected to yield similar results. These experiments included wind tunnel tests with various power plant configurations.

It is therefore not surprising that Junkers considered these results when design work on the Ju 287, project designation EF 122, began in early 1943. The experiments had revealed that a forward sweep angle of 20° (along the quarter-chord line) was the most favorable. A critical Mach number of .85 was achieved with this configuration. Further experiments revealed the following:
- the forward-swept wing had better low-speed characteristics than a conventional swept wing, since it could reach greater angles of attack before flow separation occurred.
- In the high-speed range the forward-swept wing was free of interference, since shock waves originating at the wing root were unable to reach the wingtips.
- the vertical fin had to be designed in such a way as to compensate for the poor directional stability of the forward-swept wing.

Junkers Ju 287 V1.

The ultimate wing was completed at the end of 1943. So as to be able to begin testing immediately, the Ju 287 V1 was built as an improvised test-bed (see specification below).

The Ju 287 V1 made its first flight at Brandis on 16 August 1944, taking off with the help of three Walter HWK 109-501 takeoff-assist rockets which were jettisoned thirty seconds after takeoff.

Junkers Ju 287 V1 in flight.

Ju 87 V1 Technical Description

Purpose:	four-engined jet bomber flying test bed
Crew:	two, in fully glazed nose section
Wing:	cantilever mid-wing monoplane with two-piece wing with 25° of negative sweep (equivalent to 20° of sweep on the 25% chord line). Two-spar structure with smooth metal skinning, two-part differential ailerons in the outer sections, trailing edge flaps on inner sections. Fixed slots in wing leading edge at the root.
Fuselage:	all-metal monocoque fuselage, almost rectangular in cross-section, mainly built using He 177 components, fully glazed nose section from the He 177 A-3.
Tail surfaces:	cantilever all-metal conventional tail, taken complete from the Junkers Ju 388.
Undercarriage:	fixed tricycle undercarriage. Mainwheels on shock struts under the wings, braced against outer wing. Two closely-spaced nosewheels, each on its own shock strut, taken from captured Convair B-24 Liberator. All wheels with streamlined fairings. Additional fixed tailwheel under the aft fuselage.
Power plants:	four Junkers Jumo 004 B-1 turbojets, each producing 900 kg of static thrust, two on the sides of the fuselage nose and two beneath the wing center section.
	In addition, four Walter HWK 109-502 liquid-fuel rockets for takeoff assist, one under each jet engine.

Junkers Ju 287 V1. Clearly
visible are the forward-swept
wing and the takeoff-assist rocket
mounted beneath the starboard
wing engine. In front of the
vertical stabilizer is a camera to
record wool tuft airflow trials.

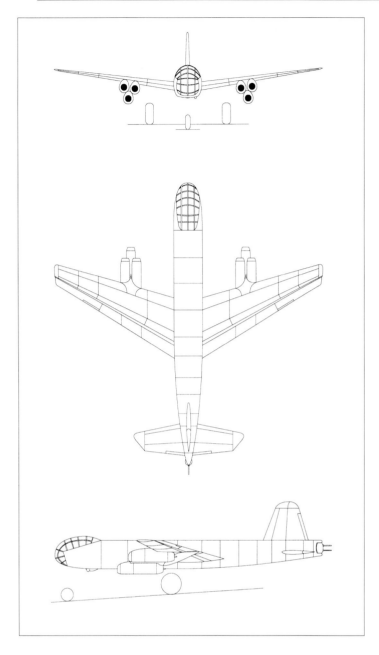

Six-engined Junkers Ju 287 A-1.

fuselage nose, and four BMW 003 turbo-jets in pairs beneath the wings. This was later changed to six BMW 003 turbojets installed in clusters of three beneath the wings. The V3 was supposed to use the fuselage of the Ju 288 with six BMW 003 power plants in a similar arrangement to the V2. The V4 was laid out like the V3. The V5 and V6 had a revised rear fuse-lage designed to accommodate two MG 131 machine-guns (FHL 131Z) and was the predecessor of the production ver-sion. The A-1 and B-1 pre-production aircraft were to be powered by six Jumo 004 C turbojets in clusters of three be-neath the wings. These aircraft were de-signed to carry an internal bomb load of 4,000 kg and 2,000 liters of fuel.

At the end of 1944 work on the Ju 287 V1 had to be suspended as a result of the Emergency Fighter Program. It was resumed at the beginning of 1945, however, and the RLM simultaneously ordered series production of the Ju 287 to go ahead. One hundred aircraft were to be manufactured by the ATG (Allge-meine Transportanlagen GmbH) in Merseburg, near Leipzig, by September 1945.

The Ju 287 was originally supposed to be powered by two Jumo 012 turbo-jets each producing 2,900 kg of static thrust, or two BMW 018 turbojets each producing 3,400 kg of static thrust. This was later examined further under project designation EF 125. Since nei-ther power plant was available, it was decided to switch to the Heinkel HeS 011 which was then under development. Four of these turbojets, each producing 1,300 kg of static thrust, were to be in-stalled in a similar arrangement to that of the V1. When this engine also became unavailable, a switch was made to six BMW 003 turbojets for the production aircraft, with three engines beneath each wing in a triangular arrangement. Wind tunnel tests revealed that this ar-rangement caused wing vibration, and furthermore, access to the power plants for maintenance was poor.

After sixteen flights the V1 was handed over to Rechlin where flow tests were carried out. The aircraft was heavily damaged in a bombing raid. The V1 was rebuilt, but it is unlikely that it flew again before the end of the war.

In mid-September 1944 the RLM had ordered five additional prototypes and two pre-production aircraft.

The V2 was supposed to be similar to the V1, except that the horizontal tail was to be lowered by 300 mm. Initially, the V2 was to be powered by two Jumo 004 B turbojets, one on each side of the

Technical Description of the Ju 287 V5, Prototype for the Production Version

Purpose:	six-engined jet bomber
Crew:	three men in a pressurized cockpit
Wing:	cantilever mid-wing monoplane, all-metal wing with 25° of negative sweep. Construction like the Ju 287 V1
Fuselage:	all-metal monocoque fuselage with almost rectangular cross-section. Fully glazed pressurized cockpit.
Tail surfaces:	cantilever conventional all-metal tail surfaces, all control surfaces equipped with trim tabs
Undercarriage:	retractable tricycle undercarriage. Mainwheels retract inwards into the fuselage, with wheels turning through 45° to lie vertically next to each other in fuselage. Twin nosewheels retract rearwards into fuselage.
Power plants:	six BMW 003 B turbojets each capable of producing 800 kg of static thrust. Installed in pairs under the wings and one on each side of the forward fuselage.
Armament:	one remotely-controlled barbette with two 13-mm MG 131 machine guns (FHL 131 Z), control by means of a periscopic sight. Bomb load of 4 500 kg in a bomb bay in the forward fuselage.

The major components of the V3 were complete at the end of the war. Together with the Ju 287 V1 and V2 (similar layout to the V1), they fell into Soviet hands in April 1945. On Soviet orders the V3 was completed, and in September 1945 it was moved to Podberezhye, 120 km north of Moscow, together with associated equipment, the technicians, and their families. There work continued in OKB 1 under Dipl.Ing. Brunolf Baade. The V3 completed 200 flying hours and had a significant impact on Soviet jet bomber development.

The EF 131 was a derivative of the Ju 297 V3, and at the end of the war it was also partially complete. On orders from the Soviet occupation forces work on the EF 131 resumed in the summer of 1945. It is believed to have been made ready to fly at Dessau in the summer of 1946. The aircraft was subsequently disassembled and shipped to Moscow, where flight testing began in the early summer of 1947.

List of Completed and Planned Prototypes

First Flight		Remarks
V1	16/08/1944	four Jumo 004 B turbojets, hybrid aircraft for testing the forward-swept wing
V2		six BMW 003 A turbojets in two clusters of three
V3		six BMW 003 A turbojets, completely reworked aircraft
V4		as V3 (project)
V5		as V3, with FHL 131 Z tail turret (project)
V6		as V5 (project)

Twin-Engined Flying Wing Jet Fighter

Horten Ho IX

The brothers Walter and Reimar Horten began experimenting with all-wing aircraft at the beginning of the 1930s. In the beginning they built gliders, followed soon afterwards by powered variants, such as the H IIm *D-Habicht*. In the process they gained extensive experience in the field of all-wing aircraft. The breakthrough came with the H III, which awoke the interest of the RLM, as a result of which Horten received a certain amount of support.

The real impetus to build the Ho IX came in 1943, however, when Walter Horten visited Messerschmitt. There he saw the Me 262 and received performance data on the Junkers Jumo 004 turbojet. The Horten brothers decided to submit a design for the 1,000 x 1,000 x 1,000* bomber demanded by the RLM.

In August 1943 Göring personally issued a contract for Horten to build an unpowered prototype within six months.

The V1 airframe was complete at the beginning of 1944, and it made its first flight on 28 February. After initial flight tests it went to the High-Altitude Research Station in Oranienburg where weapons tests were carried out. After this it went to Brandis for service trials.

Meanwhile, in Göttingen the Horten brothers had completed the V2, which was supposed to be equipped with two BMW 003 turbojets and a fully-retractable tricycle undercarriage. The nosewheel consisted of the tailwheel and retraction mechanism from a wrecked He 177 bomber, while much of the main undercarriage came from a Bf 109. The Ho

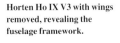

Horten Ho IX V3 with wings removed, revealing the fuselage framework.

* 1 000 km/h speed, 1 000 km range, and 1 000 kg bombload.

Horten Ho IX V1.

IX V2 was also equipped with a braking chute. Since the desired BMW power plants were unavailable, the Ho IX V2 had to be fitted with two Jumo 004 turbojets. On 2 February 1945 it took off on its maiden flight, which lasted thirty minutes. Disaster struck on the third test flight, however. On 18 February 1945 the V2 crashed and was completely destroyed.

In March 1945 the RLM decided to transfer development of the series aircraft and subsequent production to Gothaer Waggonfabrik, which had extensive experience in the building of cargo gliders and free production capacity. The aircraft was now officially designated the Go 229, although the designation Ho 229 was more commonly used. Preparations were made for the first batch of twenty aircraft. The Gothaer Waggonfabrik developed the aircraft further, and a technical description of the production aircraft appears on the following page.

The Ho 229 V3 was to be the prototype for the production version and was almost complete when the war ended. It was captured by the Americans and shipped to the United States as T2-490. The aircraft was supposed to be made flyable, however, budget cuts prevented this, and it was handed over to the National Air and Space Museum in Washington, D.C. Today it is still there, in the museum's storage facility at Silver Hill, Maryland. By now it has probably reached a pitiful condition and awaits restoration.

When the war ended, further prototypes were under development by the Gothaer Waggonfabrik:

V4 single-seat night-fighter with nose-mounted radar

V5 single-seat fighter-bomber

V6 single-seat fighter with pressurized cockpit and armament of four MK 108 cannon or two MK 103 cannon, and two Rb 50/18 cameras for the reconnaissance role; also extensive changes to the airframe.

On 1 March 1945 the Horten brothers also submitted a design for a Ho 229 V6 as a follow-on to the V3 to V5 prototypes. It was designed to fill the same roles as the Gotha designs, namely day, night, and bad weather fighter, heavy fighter, light bomber, or reconnaissance

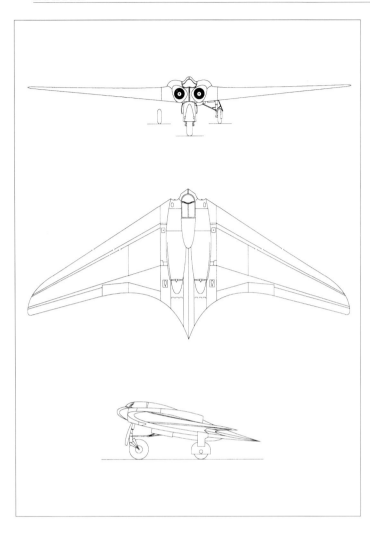

aircraft, however, it was designed as a two-seater. The pilot and observer were to sit in an armored pressurized cockpit equipped with ejection seats. The proposed armament was significantly heavier, though:

- four MK 108 cannon with 120 rounds per gun
- 24 to 36 R4M rockets
- one ETC 503 bomb rack under each engine for the carriage of 1,000-kg bombs

It was anticipated that takeoff-assist rockets would be used in this configuration. At normal weights without external loads the aircraft was expected to achieve a speed of 950 km/h at ground level.

Horten Ho IX.

Horten Ho IX Technical Description

Purpose:	single-seat, twin-engined jet fighter
Crew:	pilot in unpressurized cockpit in the nose of the center-section, special pressure suit for the pilot
Wing:	cantilever all-wing, three sections, center section framework of welded steel tube, plywood skinning except for stainless steel behind engine exhausts. Outer sections two-spar wooden construction with plywood skinning. Wingtips of light metal.
Control surfaces:	all control surfaces in wing, outer trailing edge designed as three-piece flap, larger outer section as combined aileron and elevator, two-part inner section as landing flaps. Two extendable braking flaps in rear center-section. Directional control by means of spoilers in outer wings, consisting of one narrow and one broad flap on the upper and lower side on each wingtip. All spoilers extended in low-speed flight, one spoiler segment in high-speed flight. Braking chute in fuselage tail cone.
Undercarriage:	retractable tricycle undercarriage. Large nosewheel (bearing 40% to 50% of total load) retracts rearward into center-section behind the cockpit, main-wheels inward into wing center-section.
Power plants:	two Junkers Jumo 004 B turbojets each capable of producing 900 kg of static thrust. Engines buried side by side in wing center-section. Air inlets in wing leading edge on either side of cockpit, jet nozzles on wing upper surface, 1/3 of way forward of trailing edge. Fuel in eight tanks in outer wings.
Armament:	four MK 108 cannon in wing center-section outboard of engines.

Single-Engined Jet Close-Support Aircraft

Henschel Hs 132

Henschel possessed extensive experience in building close-support aircraft, such as the Hs 129. On 18 February 1943 a specification was issued for a single-seat close-support aircraft to be powered by a conventional piston engine. After the receipt of performance data on turbojet engines from Junkers and BMW, in October 1943 this specification was changed to jet propulsion, and corresponding wind tunnel tests began in April 1944.

The Hs 132 was the first pure close-support aircraft to be designed with jet propulsion. It is obvious that Henschel adopted a layout similar to that of the He 162 for the same maintenance accessibility reasons as Heinkel, electing to place the engine on top of the fuselage. The pilot was to occupy a prone position in the glazed nose, since this position would allow him to endure greater g-forces in pull-outs and turns and make it possible to exploit the aircraft's expected extreme maneuverability. It had been determined that a pilot in the prone position could stand forces up to 12g.

Approach to the target was to be made in a shallow dive at a speed of 910 km/h. The bomb would be released automatically over the target, after which the aircraft would go into a shallow climb. All of this was to be controlled by a simple computer.

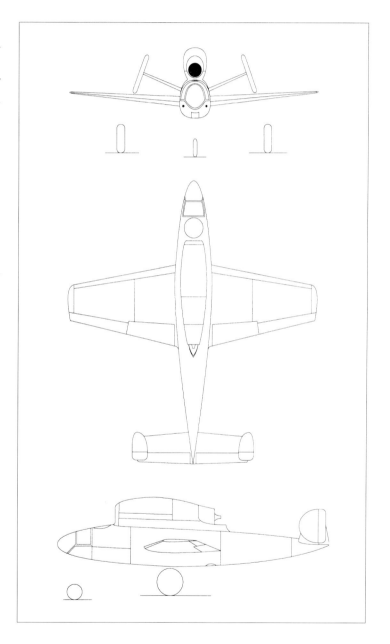

Henschel Hs 132.

71

Four different variants were projected, with the A-series to serve as prototypes:

Hs 132 A   standard close-support and dive-bomber version with BMW 003 E engine, 500 kg bomb load.

Hs 132 B   more powerful Jumo 004 B engine, bomb load of 500 kg, and armament of two MG 151/20 cannon in fuselage nose.

Hs 132 C   Heinkel HeS 011A turbojet, bomb load of 1,000 kg, and armament of two MG 151/20 and two MK 108 cannon in fuselage nose.

Hs 132 D   dive-bomber derivative of the A-series with wing span increased to 9.1 meters (wing area 16 m^2).

By the end of the war four A-series prototypes were under construction, of which the V1 was almost complete. It was supposed to make its first flight in June 1945. In March 1945 the prototype fell into Soviet hands undamaged.

The following specifications are applicable to the Hs 132 A.

Hs 132 A Technical Description

Purpose: single-engined jet close-support aircraft and dive-bomber

Crew: pilot in prone position in fuselage nose, glazed nose with glazed access hatch on top of fuselage

Wing: two part single-spar wooden wing, all plywood skinning. Control surfaces along entire trailing edge, outer part split ailerons with trim tabs, inner landing flaps

Fuselage: all-metal monocoque fuselage with nearly circular cross-section, glazed nose cone.

Tail surfaces: cantilever tail surfaces, vee-shape horizontal stabilizer and elevators, twin fins and rudders. Construction entirely of wood. All control surfaces with mechanical trimming and mass balances.

Undercarriage: retractable tricycle undercarriage, wide-track mainwheels on cantilever oleos retracting inwards into wings, nosewheel retracted hydraulically rearwards into fuselage nose.

Power plant: one BMW 003 turbojet capable of producing 800 kg of static thrust, installed on fuselage spine.

Armament: external bomb load of up to 1 000 kg beneath fuselage. With aid of takeoff-assist rockets, load increased to 1 400 kg for dive-bombing role.

Delta-Wing Interceptor Fighter

Lippisch P 13a/DM 1

Alexander Lippisch began his work on tailless aircraft in 1930. It was he who developed the Me 163 rocket fighter for Messerschmitt in which Heini Dittmar became the first person to fly faster than 1,000 km/h on 2 October 1941.

In 1943 Lippisch became head of the *Luftfahrtforschungsanstalt Wien* (Vienna Aviation Research Institute), and there concerned himself with supersonic tailless aircraft. This work resulted in the P 12 and P 13 fighter projects which took shape in 1943-44. In 1944 the RLM established the Emergency Fighter Program. Lippisch reworked his P 13 design and submitted it to the RLM as the P 13a. It was a tailless aircraft which consisted mainly of wing and power plant. Lippisch called this configuration a "powered wing."

Free flights with a scale model began in May 1944 on the Spitzerberg,

near Vienna. Wind tunnel tests began in the supersonic wind tunnel of the Aerodynamic Research Institute (AVA) in Göttingen in August 1944.

To further investigate the type's flight characteristics, the Aviation Specialists Group (FFG) of the Darmstadt Technical College, in cooperation with Lippisch, began constructing an unpowered 1 : 1 scale model designated the D 33. The Darmstadt group was bombed out on 11-12 September 1944, however, and construction of the D 33 was moved to Prien an den Chiemsee, where the Munich FFG had a hangar. With the assistance of the Munich FFG, construction was resumed on the project, which was now designated the DM 1 (Darmstadt München 1), but without the participation of Lippisch. In contrast to the P 13a, the DM 1 had a fixed tricycle undercarriage, and for trim purposes 35 liters of water could be hand-pumped back

DM 1 experimental glider, built by the Darmstadt and Munich Aviation Specialist Groups.

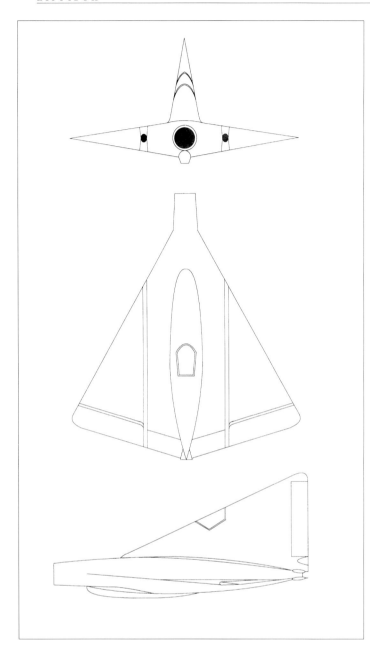

Lippisch Li P 13a delta-wing fighter with ramjet engine.

and forth between tanks in the nose and tail.

Flight testing was supposed to start at the beginning of 1945. In January 1945 two Si 204 A light transports were made available for the purpose of carrying the DM 1 aloft piggyback fashion. Once the necessary altitude had been reached it would be released. The DM 1 was then supposed to reach speeds on the order of 800 km/h in a shallow dive with the assistance of solid-fuel rockets.

On 3 May 1945 American troops occupied Prien airfield and discovered the half-completed DM 1. On 9 May General Patton inspected the DM 1 and ordered its completion. The Americans also planned to test the DM 1 using the piggyback method, for which purpose a C-47 was made available. But then orders came to transport the DM 1 to the United States and conduct the trials there. On 9 November 1945, packed in a sea crate, the DM 1 left Prien. It was transported overland to Rotterdam, and from there it was shipped to Boston.

The DM 1's destination was the full-scale wind tunnel at Langley Field in Virginia. There it was put through an extensive test program by the NACA (National Advisory Committee for Aeronautics). In the course of these trials, sharp edges were fitted on the formerly rounded leading edges in order to improve lift and flow behavior. The DM 1 was converted several times, almost to the point of be-

Lippisch P-13 Technical Description

Purpose:	interceptor fighter
Crew:	one pilot, cockpit integrated into forward area of fin
Wing:	cantilever delta wing with 60° of sweep on leading edge and 15% profile thickness, rounded wingtips; wood and steel frame with plywood skinning
Fuselage:	no fuselage in conventional sense
Tail surfaces:	partly glazed, set on wing center-section, triangular fin with 60° of sweep on leading edge and 17.5% profile thickness, rounded tip; wood and steel frame with plywood skinning
Undercarriage:	central retractable landing skid
Power plant:	ramjet engine in wing center-section with flaps in exhaust stream to supplant directional control. Planned fuel a mixture of coal dust and heavy oil. Additional rocket engine to enable aircraft to reach ramjet start speed (150 km/h).
Armament:	two MK 108

ing unrecognizable. The test program was ended in late 1947, and in January 1948 the DM 1 was placed in storage.

In 1949 the present-day National Air and Space Museum (NASM) in Washington became interested in the DM 1. In 1950 the aircraft was given to the museum, which placed it in storage at Silver Hill. It is probably still there today, awaiting restoration. Because of its all-wood construction, the DM 1 must be in deplorable condition by now.

Convair was the only American aircraft company to take an interest in the test results achieved by the DM 1. These results apparently flowed into the development of a series of delta wing projects, the XF2Y-1 Sea Dart, XF-92, F-102 Delta Dagger, and F-106 Delta Dart.

DM 1 at Prien am Chiemsee in 1945.

75

Single Seat Fighter with T-Tail

Focke-Wulf Ta 183 *Huckebein*

In March 1943 Focke-Wulf in Bremen began various design studies for a turbojet-powered fighter aircraft. Of the numerous designs, three were selected for mockup construction. These were Design 2 of June 1943, Design 5 of January 1944, and Design 6 of February 1944. These designs were subsequently renamed:

Project III (Design 2)
one Jumo 004 turbojet beneath the fuselage

Project VI (Design 5)
one turbojet engine in the fuselage integrated with takeoff-assist rockets below the rudder

Project VII (Design 6)
one HeS 011 A turbojet and one Walter HWK 109-509 A-2 rocket motor, dubbed *Flitzer.*

In mid-1944 the *Oberkommando der Luftwaffe* issued a call for tenders for a single-seat fighter to be powered by one HeS 011 turbojet engine. The proposed fighter would have an armament of two or four MK 108 cannon and would be capable of reaching a maximum speed of 1,000 km/h at an altitude of 7,000 meters (the same specification on which the Messerschmitt P1101 was based).

Focke-Wulf chose Fighter Project VI for its entry, and under the direction of Dipl.Ing. Hans Multhopp it was reworked in preparation for a conference to be held on 19 December 1944 to compare the competing designs. Project VI had a fuel tank in the fuselage and a group of six tanks in each wing. Because of the wing sweep angle of 40 degrees, these wing tanks had to be drained in a specific order to avoid cg problems.

A total of three versions was planned:

Ta 183 Technical Description	
Purpose:	single-seat jet fighter
Crew:	one pilot in pressurized cockpit with ejection seat
Wing:	cantilever mid-wing monoplane, two-part single-spar wooden wing with 40° of sweep (32° degrees along the quarter-chord line). Landing flaps between ailerons and fuselage.
Fuselage:	all-metal monocoque fuselage with raised tail bearer flowing into vertical stabilizer
Tail surfaces:	cantilever all-metal conventional tail, horizontal tail of shallow vee shape mounted atop vertical stabilizer. Horizontal stabilizer swept 40° along the quarter-chord line.
Undercarriage:	retractable tricycle undercarriage. Mainwheels retract forward into fuselage sides, nosewheel rearward into fuselage nose.
Power plant:	one Heinkel HeS 011 turbojet capable of producing 1 300 kg of static thrust. Central air intake in fuselage nose with direct flow to engine.
Armament:	two (four) MK 108 in fuselage nose

- interceptor fighter
- fighter-bomber
- reconnaissance aircraft

Preliminary work on the project had already reached an advanced stage, and Focke-Wulf received a development contract for its submission, which was given the designation Ta 183 *Huckebein*.

Design work began in January 1945, and all the necessary drawings had been completed by the end of the war. It had already been determined that the first production version of the Ta 183 would be powered by the Jumo 004 B turbojet producing 900 kg of thrust, as the Heinkel HeS 011 was not yet available in sufficient numbers.

A total of 16 prototypes was ordered. The V1 to V3 prototypes were to be powered by the Jumo 004 B turbojet, and the V4 to V14 by the Heinkel HeS 011 A. Prototypes V15 and V16 were earmarked for static testing. Of course, none of these prototypes was built.

The Ta 183's design layout with its single turbojet contained in the fuselage was strongly reminiscent of the Messerschmitt P1101. These two projects were probably the most advanced jet aircraft projects of their time, and both influenced jet fighter development in east and west for many years after the war.

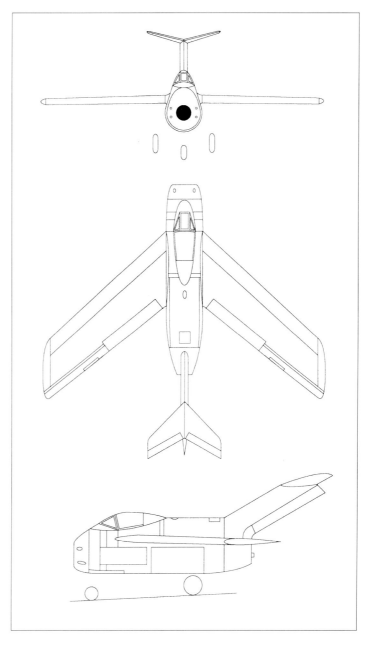

Focke-Wulf Ta 183.

Turbojet Engines

Apart from the private research and development work promoted or carried out by Ernst Heinkel, official work on turbojet engines under the direction of the RLM began at the end of 1938.

At the end of August 1938 in Munich there was a special meeting of the Lilienthal-Gesellschaft titled "Airframes and Power Plants." Prof. Karl Leist of the DVL gave a speech on the theme, "The Gas Turbine as the Main Power Plant for Aircraft." He listed the advantages of this type of propulsion, including:

- the absence of moving masses
- greater life through uniform loads on the system
- ability to use cheaper domestic fuels
- reduced frontal drag

The DVL had already been researching various special power plants since 1936. In two of these thrust was produced by a propeller driven by a gas turbine. Turbojet engines were also discussed, but only in general terms. However, this was sufficient for Helmut Schelp of the RLM (Director of Special Power Plants) to begin a campaign for the development of turbojet engines, even though the aviation industry at first showed little interest.

The first development program for turbojet engines was begun at the end of 1939 and included six power plants:

109-001	Heinkel HeS 8 (Dr. von Ohain)
109-002	Weirich jet engine by BMW (Dr. Oestrich)
109-003	BMW 003 (Dr. Oestrich)
109-004	Junkers Jumo 004 (Dr. Franz)
109-006	Heinkel HeS 30 (M.A. Müller)
109-007	Daimler Benz ZTL (Prof. Leist)

Drawing from the code 9 for aeroengines, the RLM assigned the code 109 to jet engines.

On 31 January 1941 a sitting of the German Academy for Aviation Research in Berlin held a working day on the theme "Turbojet Engines," during which twelve addresses were given on the actual state of research and development. In his speech Helmut Schelp gave a breakdown of the various types of engine which is still valid today:

TL (Turbojet) Power Plant: Air drawn into the air intake is compressed in the compressor. The compressed air is then heated in the combustion chamber. In the turbine the hot gas expands somewhat, releasing sufficient energy for the turbine to drive the compressor. The energy potential of the gas is not yet exhausted, however. In the jet nozzle the thermodynamic energy stored in the gas is converted to kinetic energy. The result is a high exit speed, which is a requirement for the production of thrust.

PTL (Turboprop) Power Plant: If the gas turbine is designed in such a way that its output exceeds the requirements of the compressor, the excess output can be used to drive a propeller. The jet thrust of the turboprop is less than that of the turbojet, because the turbine takes more energy from the gas stream. The principal element of the turboprop does not differ from the turbojet, however, additional reduction gearing is required for the propeller.

ZTL (Turbofan) Power Plant: Like the turboprop, the turbofan engine produces more output than is required by the compressor. This excess output is used to drive a low-pressure compressor, which produces a second air stream in an outer channel (bypass channel) that surrounds the actual power plant. Most of the expansion takes place in a common jet nozzle. In particular, turbofan engines are more effective than turbojet or turboprop engines in the 900 km/h speed range.

ML Power Plant: Two-stage engine in which the low-pressure compressor is driven by a motor.

Influenced by the war situation, in 1942 there was a thorough rationalization of the program which resulted in all ML projects, designs with centrifugal-flow compressors, and power plants with complicated contra-rotating compressors being canceled or reduced in priority. The plan was as follows:
Series production:
    BMW 003 and Jumo 004
Development:
    Heinkel HeS 011
    (second-generation power plant)
Preliminary development:
    BMW 018 TL and 028 PTL
    Heinkel / DB 021 PTL
    Jumo 012 TL and 022 PTL
    In the years that followed, however, the Emergency Fighter Program and the mission of home defense meant that only the first two stages, production and development, could be carried out. Nevertheless, on 20 July 1944 the RLM issued the following development plan for special power plants:

109-003A
TL engine, 800 kg static thrust, for Ar 234 C, Me 262 and Ju 287

109-003R
TLR engine consisting of the 003A and rocket propulsion,
800 + 1,250 kg static thrust, for Me 262 interceptor

109-004B
TL engine, static thrust 900 kg, for Me 262 and Ar 234 B

109-011A
TL engine, 1,300 kg static thrust, For Ar 234, Ju 287 and He 343, performance class III

109-012
TL engine, 2,780 kg static thrust, For Ju 287 and He 343, performance class III

109-018
TL engine, 3,400 kg static thrust, for high-altitude fighter and high-altitude reconnaissance aircraft, performance class IV

109-021
PTL engine, equivalent shaft output horsepower 2,400, for jet bomber project, performance class II

109-022
PTL engine, equivalent shaft output horsepower 4,600, for jet bomber project, performance class III

109-028
PTL engine, equivalent shaft horsepower 7,570, for medium-range bomber, performance class IV

Heinkel Jet Engines
Pabst von Ohain concerned himself with the theoretical design of jet engines even while a student, and later as an as-

Cross-section of the Heinkel
HeS 3 B engine.

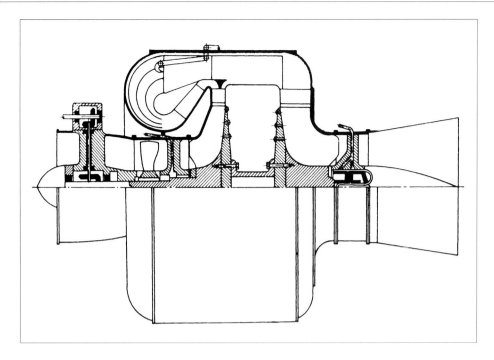

sistant at the Physical Institute of the University of Göttingen under Prof. Pohl. In 1934 he met Max Hahn, who worked in the auto repair garage where he took his car. In him he found the technical assistant who could check his designs for technical feasibility and then turn them into reality.

Pabst von Ohain designed an experimental turbine with a centrifugal-flow compressor which was built by Max Hahn. During test runs, however, it proved impossible to get the turbine to run on its own. Motive power was provided by a grinding machine from the garage. It was possible to obtain pressure and temperature readings, and these results provided a valuable basis for subsequent work.

In spring 1936 Prof. Pohl arranged a meeting between Pabst von Ohain and Dr. Ernst Heinkel. The latter immediately grasped the enormous potential of an aircraft with jet propulsion. As a result, Ohain and Hahn were hired by Heinkel. A department called Special Developments II was founded in the Heinkel works in Rostock-Marienehe. Maximum secrecy was maintained, and it was

sealed off from the rest of the works. Even the RLM remained unaware of Heinkel's activities. At the same time, Heinkel gave his airframe designers Siegfried and Walter Günther the task of designing an airframe to accept the jet power plant, the He 178.

In April 1936 Ohain and Hahn met in Rostock-Marienehe, where they were joined by Dipl.Ing. Wilhelm Gundermann. The first thing they did was investigate the failure of the first test run at Göttingen, and after much testing they discovered the reasons:

- turbine wheel
- absence of guide vanes
- insufficiently large combustion chamber

When the first indications of success appeared, the team was bolstered. Ohain was allowed to select the people he wanted from the various departments of the Heinkel works. In February 1937 the demonstrator turbine HeS 2 was completed, and it made its first bench run at the beginning of March, developing 80 kg of static thrust. In subsequent runs it reached its calculated static thrust of 125 kg. Water had to be used to

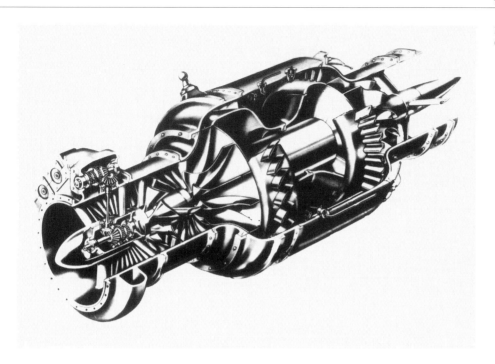

power the turbine, however, since there were still problems with injecting fuel into the combustion chamber. Nevertheless, the turbine could be demonstrated, which was of significant importance.

Further burn attempts were now made in a wind tunnel. When satisfactory results were achieved, efforts were directed at the first aircraft power plant, the HeS3. It was to have an axial first stage with eight vanes made of Dural, followed by a single-stage centrifugal-flow compressor with sixteen vanes made of heat-resistant Duraluminum W. The air stream was redirected and split into three streams: one to cool the combustion chamber, one to cool the mixture chamber, and one for burning in the combustion chamber. The sixteen individual combustion chambers were made of heat-resistant steel with a 36% nickel content. The following turbine with sixteen vanes was made of a Krupp special material, since turbine steel was not available. During one of the first bench runs in spring 1938 the HeS 3A developed 450 kg of static thrust.

Since the He 178 airframe was still incomplete, the power plant was in-stalled beneath the fuselage of an He 118 for test flights. The He 118 achieved a clear increase in speed after the engine was turned on. However, the HeS 3A was completely destroyed in a fire caused by the ignition of residual fuel after landing. As a result, when the He 178 airframe was completed, it was necessary to install the little tested but more powerful (500 kg static thrust) HeS 3B in the aircraft. On 27 August 1939 the He 178 made the first flight by a jet aircraft anywhere in the world.

Even though this success received little recognition from the RLM, in October 1939 Heinkel began setting up a larger power plant section in Rostock-Marienehe.

As an interim step, Heinkel developed the HeS 6 engine from the HeS 3B. Of the same dimensions, but with a more efficient compressor, it produced 590 kg of static thrust. Meanwhile, Heinkel had initiated a twin-engined jet fighter project on which engine dimensions had a decisive influence. This eliminated engines such as the HeS 6 with return-flow combustion chambers, which were too large. It was therefore decided to devel-

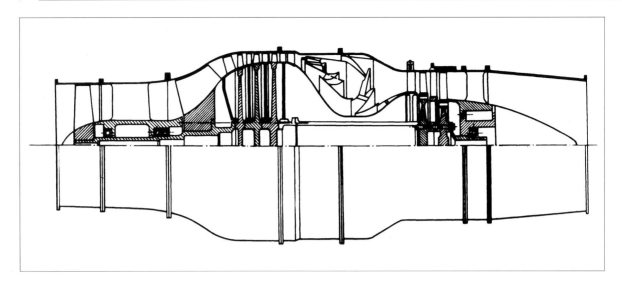

Heinkel HeS 011 engine.

op a completely new engine designated the HeS 8. It would have straight-flow combustion chambers, reducing engine diameter from 1,200 mm to 775 mm. The HeS 8B turbojet developed 750 kg of thrust and weighed just 390 kg. Its first live use was the maiden flight of the He 280 on 2 April 1941.

In parallel to the HeS 8, Heinkel developed the axial-flow HeS 30 turbojet engine. This engine developed 750 kg of static thrust in its first test runs, and later 820 kg. At this time the more powerful Junkers Jumo 004 turbojet was preparing to go into quantity production, and it had been announced that the BMW 003, which was in the same thrust category, was production-ready. Consequently, the RLM shelved development of the HeS 30 engine, even though it was lighter and smaller in diameter than its contemporaries.

In the opinion of those responsible in the RLM, axial compressor engines had greater thrust potential than those with centrifugal compressors, therefore, Heinkel was unable to obtain a production contract for his HeS 8 engine. He did have an advocate in Udet, however, and in April 1941 he was able to purchase the Hirth aero-engine factory in Stuttgart-Zuffenhausen. This increased

capacity would enable Heinkel to expand its power plant activities.

At the very latest, the skeptics in the RLM had their last doubts removed when representatives of the RLM witnessed a demonstration of the He 280, in which it easily bested a Fw 190 in a mock dogfight. The advantages of jet propulsion for aircraft and the unique opportunity which it offered Germany were now obvious to everyone involved.

As a result, in September 1942 Heinkel received a development contract for a more powerful turbojet engine which was to power a new, faster fighter. The RLM assigned the designation 109-011 to this engine, while Heinkel called it the HeS 011. The RLM specification for the 109-011 power plant called for a static thrust of 1,300 kg, a maximum weight of 900 kg, and a maximum power altitude of up to 15,000 meters. In addition, the engine was to have the potential of producing 1,500 kg of static thrust.

Preliminary design work began in the late summer of 1942, with Heinkel designers working closely with the RLM. In order to test the selected configuration, five experimental engines were built without regard to weight or future production considerations.

The engine's compressor consisted of an axial front stage, an oblique-flow diagonal stage, and three axial stages. The compressor's flow volume was to be 30 kg/s at a speed of 11,000 rpm. The combustion chamber was designed as an annular unit, which allowed the minimum possible distance between the compressor and turbine. In spite of known problems, such as those encountered during start-up, the first experimental models were fitted with an annular combustion chamber based on that of the HeS 8. At the same time design was begun of a new sixteen-nozzle chamber for high pressure injection. A two-stage turbine was selected in order to effectively take advantage of the high heat gradient. The first test systems were fitted with massive uncooled turbine blades, whereas later production versions were to have air-cooled hollow blades. As usual, regulation of the power plant was by selecting rpm. But unlike all other engines, one jet nozzle setting was to be used for all flight situations. A second nozzle setting was planned only for starting the engine and for high idle while on approach to land.

Testing of various compressor arrangements began in May 1943, and construction of the first complete test engine, the V1, was completed in September 1943. During test runs at overload, however, this power plant produced just 1,100 kg of static thrust. The principal reason for this was a flaw in the design of the axial stage of the compressor: there were problems with the turbine's solid blades, as well as unfavorable temperature distribution in the combustion chamber and excessively high pressure loss. The necessary modifications were made during construction of the prototype engines V6 to V25. Engine V6 completed its first test run in February 1944, but it still was not fitted with the totally optimized compressor. This was completed in May 1944, and in June was installed on the V6. During the course of a test run lasting several hours the engine reached static thrust levels of 1,000 to 1,100 kg. Work had not yet begun on the combustion chamber problem. Development work on the sixteen-nozzle chamber was completed in November 1944. Static thrust levels of 1,200 to 1,300 kg were reached in test runs in December 1944.

Design work on the pre-production HeS 011 A-0 power plant began in May 1944 and was completed in December. It had originally been planned to build additional prototypes, the V26 to V85, but these were canceled at the urging of the RLM. Instead, several of the first A-0 pre-production engines were to be used for testing. Delivery of the first pre-production engines was expected at the end of May 1945, but because of the relocation of the Hirth works and several subcontractors in April 1945 this was not possible.

A total of twenty HeS 011 power plants were manufactured. When the Americans occupied the factory in Stuttgart they gave orders for the last eight HeS 011 A-0 pre-production engines to be completed. They were transported to America, where they were extensively bench-tested by the U.S. Navy in Trenton.

In designing this engine Heinkel stressed extreme simplicity and ease of manufacture (with the minimum possible build time and the ability to use unskilled labor).

83

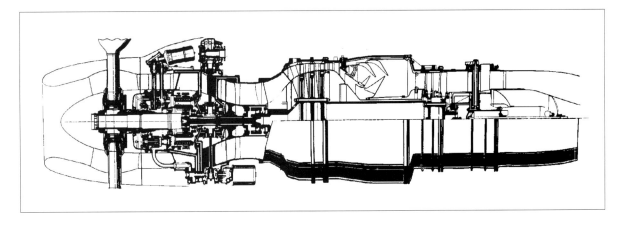

HeS 21 turboprop engine based on the He 5011.

As well as turbojet engines, Heinkel was involved in other projects using gas turbines:

HeS 9
Turboprop engine based on the HeS 8

HeS 21
Turboprop engine based on the HeS 011

HeS 10
Turbofan engine based on the HeS 8; developed 740 kg static thrust; three prototypes built in 1939-40.

HeS 30 A-ZTL
Turbofan engine producing 1,050 kg static thrust.

HeS 60
MTL engine; combination of TL and ML power plants; more favorable fuel consumption at cruise setting, and three times more thrust.

All of these projects had to be abandoned in 1942 in favor of development of the HeS 011 engine.

Junkers Jet Engines

Junkers began top-secret work on a "gas turbine" in 1927. This came to an abrupt end in 1933 when the National-Socialists nationalized the Junkers company.

Overview of Heinkel Jet Engines						
Designation	HeS 3B	HeS 6	HeS 8	HeS 10	HeS 11	HeS 30
Year	1939	1939	1941	1940	1944	1941
Type	TL	TL	TL	ZTL	TL	TL
Compressor	1A+1R	1A+1R	1A+1R	2A+1R	1A+1D+3A	5A
Combustion						
Chamber	annular	annular	annular	annular	annular	11
Turbine	1R	1R	1A	2A	2A	1A
Dimensions						
Length mm	1 630	1 700	1 675	4 650	3 345	2 850
Diameter mm	1 200	1 200	775	1 640	805	562
Dry weight kg	360	420	386	500	885	375
Static thrust kg	500	590	720	740	1 300	750
RPM	13,000	13,300	13,500	13,500	11,000	10,500
Specific fuel						
consumption						
kg per km/h	1.60				1.30	1.19
RLM Designation			109-001		109-011A	109-005
Remarks	flown in He 178	flown under He 111	flown in He 280	experimental engine test stand	quantity production planned, test stand	experimental engine, test stand

Junkers Jumo 004 engine

Work on an aircraft turbine was resumed in 1935. In 1936 and 1937 Junkers worked with the AVA Göttingen and built a multi-stage axial compressor for experimental purposes. Furthermore, in the intervening period much experience had been gained with gas-driven turbosuperchargers, which had been developed for the L88 and Jumo 205 piston engines.

In 1937 at Junkers' Magdeburg branch plant, a team led by Ing. Max A. Müller began designing and developing a turbojet engine with an axial compressor. Various preliminary experiments were conducted to investigate axial compressor configurations and turbine blade profiles, as well as burning experiments. The experimental gas turbine is said to have run independently at the beginning of 1939, but when problems arose with the allocation of special materials, for example, heat-resistant steel, the RLM ordered the project

turned over to an engine manufacturer. Prof. Wagner was reluctant to give up control of the project. He finally returned to the Berlin Aviation Technical Institute, and Ing. Max A. Müller went to Heinkel.

The RLM subsequently issued a contract to the Junkers company in Dessau to develop a turbojet engine which was to produce 600 kg of static thrust and have a maximum diameter of 600 mm. Dr. Anselm Franz was responsible for this project.

Design work on the Junkers T-1 engine began in 1939. The power plant was to meet the following parameters:

Thrust at altitude of 1,500 m	
at 900 km/h	600 kg
Static thrust	900 kg
Flow rate	20 kg/s
Compression ratio	3.2
Maximum rpm	9,000

The resulting design was a single-shaft engine with an eight-stage axial

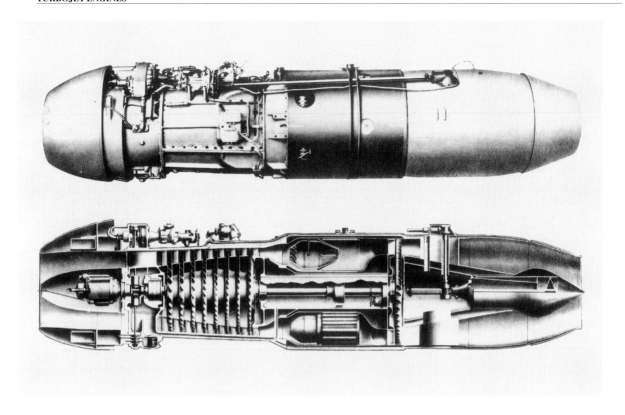

Junkers Jumo 004 engine

compressor (based on the work with the AVA Göttingen and experience with exhaust-driven turbosuperchargers), six individual combustion chambers, and a one-stage axial turbine, in whose development the AEG company participated. The T-1 engine made its first 30-minute test run on 11 October 1940. Maximum speed of 9,000 rpm was reached for the first time in December 1940. In January 1941 a static thrust of 450 kg was measured, and on 6 August 1941 during a test run the engine achieved the specified static thrust of 600 kg.

The T-1 completed its first ten-hour run on 24 December 1940, briefly reaching an overload of 1,000 kg of static thrust. During this period the engine was subjected to a whole series of changes, improvements, and redesigns. On 15 March 1942 the T-1 began flight trials mounted beneath a Bf 110.

The breakthrough with the RLM came after Junkers delivered the V9 and V10 prototypes to Messerschmitt, which

used them to propel the Me 262 V3 on its first pure-jet flight on 18 July 1942. The RLM ordered the construction of 80 power plants for test purposes. The engine's official designation was 109-004A. Quantity production of the Jumo 004 A-0 started at the beginning of 1943. In February 1943 two of the first engines, serial numbers 006 and 016, were installed in the Ar 234 V1, which made its first flight with them on 15 June 1943.

The A-0 series was designed and built without concern for rationed materials like nickel, cobalt, chrome, and molybdenum. By mid-1942 it was already clear that the engine could never be built in quantity using the current design. For example, a single Jumo 004 A-0 engine used 88 kg of nickel. Junkers began seeking and evaluating alternatives in the summer of 1942. The rationed materials problem made it necessary to undertake a radical redesign of the power plant, resulting in the Jumo

Overview of Junkers Engines

Designation	004 B	004 D	004 H	012	022
Year	1941 (1943)	1944	1945	1945	
Type	TL	TL	TL	TL	PTL
Compressor	8 A	8 A	11 A	11 A	11 A
Combustion chambers	6	6	8	8	8
Turbine	1 A	1 A	2 A	2 A	3 A
Dimensions					
Length mm	3 864	3 864	3 950	4 945	5 640
Diameter mm	765	765	860	1 080	1 080
Dry weight kg	850		1 100	1 600	2 600
Static thrust kg	910	930	1 800	3 000	4 600
					(equivalent shaft H.P.)
RPM	8,700			6,600	5,300	5,300
Specific fuel consumption					
kg per km/h	1.40		1.20	1.20	
RLM designation	109-004			109-012	109-022
Remarks	quantity production	production	design	design	development based on 012

004 B. Almost every component of the engine which was exposed to heat was now made from FLW 1010 stamped sheet metal. Special precautions had to be taken to cool the components, and the outer surfaces were coated with aluminum to prevent oxidation. The first Jumo 004 B-0 completed a bench run in May 1943. The B-0 engine was 100 kg lighter than the A-0. Tinidur, an austenized steel manufactured by Krupp, was initially used for the turbine blades. This steel was suitable to 720 degrees and contained only 30% nickel (instead of 80%), 15% chrome (instead of 30% previously), and 1.7% titanium. This steel was very brittle, however, which led to manufacturing problems. The designers instead turned to Cromadur, a steel alloy which contained 13% chrome, 18% manganese, and 0.7% vanadium, and was nickel-free.

In the beginning the B-series was plagued by turbine blade failures. It was finally found that the individual turbine blades had very different natural resonance frequencies, which could lead to failures under the influence of the combustion chambers. The design of the turbine blades was changed to raise their resonance frequency, and the engine's maximum revolutions dropped from 9,000 to 8,700. Meanwhile, the hot elements had reached a life of 50 operating hours, even though engines were rarely used for more than 25 hours in operational aircraft before they were changed. Development was slow, and just 626 Jumo 004 B engines were delivered by August 1944. Deliveries of Jumo 004 B engines to Messerschmitt began in February 1944. A total of 6,010 Jumo 004 B engines were built by the various manufacturing facilities in the period from February 1944 to March 1945. In spite of these impressive production figures, problems in operating the engine above an altitude of 4,000 meters persisted until the end of the war. The engine control was designed as a single lever operation. The pilot selected the power setting and had to constantly monitor the gas temperature. The indirect governor protected the engine against mechanical and thermal damage. From the beginning the Jumo 004 engine was designed to burn diesel oil (later designation J2), and consequently the RLM's or-

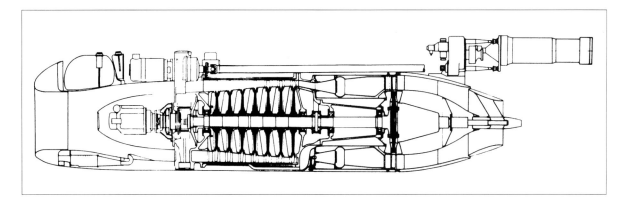

BMW 003 R with rocket booster engine.

der to switch turbines from aviation gasoline to J2 fuel caused Junkers no problems, unlike BMW. The Jumo 004 B was installed in production versions of the Me 262 and Ar 234 B.

At the end of the war Junkers had several variants of the Jumo 004, and several entirely new engines under development or about to enter production:

004 C

1,200 kg static thrust with afterburning; did not enter production.

004 D

improved air intake compared to the 004 B; 930 kg static thrust; a very few production engines constructed by the end of the war.

004 E

increase in turbine inlet temperature to 870 degrees, resulting in output of 1,000 kg static thrust, with afterburning 1,200 kg static thrust for 15 sec.; ready for production in spring of 1945.

004 H

11-stage compressor and 2-stage turbine; 1,800 kg static thrust; project only.

012

turbojet engine with output of 3,000 kg static thrust; some components complete by the end of the war; alternative to the BMW 018 for the Ju 287.

022

turboprop engine with output of 4 600 equivalent shaft horsepower

BMW Jet Engines

At Bramo it was principally Dr. Oestrich who was interested in jet propulsion, this having begun in 1928 when he was still with the DVL (German Aviation Research Institute). In 1938-39 BMW was working on Project F9225, a turbojet engine with a seven-stage axial compressor and two-stage turbine.

At the beginning of 1939 the RLM issued a development contract to Bramo and Junkers for a turbojet engine with a maximum diameter of 600 mm capable of producing 600 kg of static thrust.

In 1939 Bramo was taken over by BMW, and in September of that year all engine-related activities were concentrated in the former Bramo works in Berlin-Spandau.

Reacting to the RLM contract, in spring 1939 Bramo began development of Project P3302, ten prototypes of which were to be built. But Bramo, and, after the merger, BMW, lacked experience in the field of jet engines, and countless problems and difficulties were encountered.

On the recommendation of the AVA Göttingen, a six-stage axial compressor was chosen. Lack of knowledge in the field meant that a great deal of combustion chamber experiments were necessary. Both Bramo and BMW did, however, have extensive experience in the area of exhaust-driven turbosuperchargers,

and consequently cooled, hollow compressor blades were chosen. Design work was completed during 1940, and construction of the prototypes began. On 20 February 1941 the first prototype made its first test run. Only 150 kg of thrust was produced. The engine ran independently, but serious problems soon developed:

- tears in the weld seams between the turbine disc and turbine rotor blades
- damage to the turbine rotor blades at 8,000 rpm (design speed 9,000 rpm)
- uneven temperature distribution in the combustion chamber

In an effort to increase thrust, the number of compressor stages was increased from six to seven. At the urging of Prof. Messerschmitt, in the summer of 1941 a Bf 110 was made available for test flights. Messerschmitt expected the BMW engine to become available in 1942, and he had a not entirely selfless interest in pushing the development of the P3303. Meanwhile, the engine reached 450 kg of static thrust on the test bench. It was clear, however, that this design would never reach the required 600 kg of static thrust.

It was decided to completely redesign the V11 to V14 prototypes. Among the modifications were:

- increase in rpm from 9,000 to 9,500, resulting in a 30% higher flow rate
- 20 mm increase in turbine diameter
- resulting increase in engine diameter from 600 to 690 mm
- first use of thrust nozzle with axially-adjustable exhaust cone
- lengthening of the combustion chamber by 150 mm
- modification of the fuel injection process (formerly baffle plates)
- turbine rotor blades attached to the turbine disc by cylindrical rods instead of being welded on (blade life 40 to 50 hours)

The first bench runs of the redesigned engine took place at the end of 1941. At the beginning of March 1942 the V11 and V14 prototypes were delivered to Messerschmitt, where they were installed in the Me 262 V1. Shortly after the aircraft took off, however, both engines suffered turbine failures.

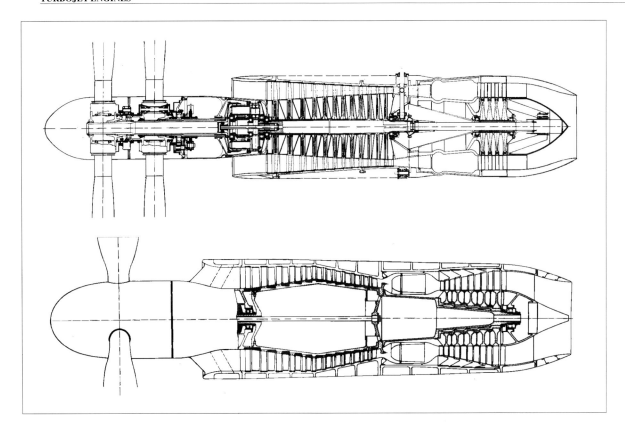

BMW 028 turboprop engine.

BMW managed to get a handle on the problems, however, and in 1943 construction of the pre-production A-0 series of the engine, now designated the BMW 003, could begin. The pre-production engines employed new turbine blades developed by BMW and built by WMF. On the test bench the engine now developed 800 kg of static thrust and 920 kg for brief periods (30 seconds). Further flight tests were carried out with the engine mounted beneath the fuselage of a Ju 88. Delivery of the last 100 pre-production engines took place in Au-

Overview of BMW Jet Engines				
Designation	003 A	003 C	018	028
Year	1943	1944	1944-45	1944
Type	TL	TL	TL	PTL
Compressor	7 A	7 A	12 A	12 A
Combustion chamber	annular	annular	annular	annular
Turbine	1 A	1 A	3 A	4 A
Dimensions				
Length mm	3 530	3 530	4 750	6 000
Diameter mm	690	690	1 250	1 250
Dry weight kg	570		2 200	3 600
Static thrust kg	800	900	3 400	6 570 equivalent shaft hp
RPM	9,500	9,800	5,000	5,000
Specific fuel consumption				
kg per km/h	1.40	1.27	1.18	
RLM designation	109-003 A		109-018	109-028
Remarks	series		production	design
	A for Ar 234			development
	E for He 162			

gust 1944. On 1 February 1944 the Ar 234 V8 had made its maiden flight powered by four BMW 003 A-0 engines.

The engine was started by means of a Riedel starter, a highly-developed two-cycle horizontally-opposed engine. Regulating engine speed in flight was very simple. The pilot selected an rpm, which the direct governor kept constant by regulating fuel flow. In autumn 1943 the RLM ordered that all turbojet engines use J2 fuel (similar to diesel) instead of aviation gasoline. A different governor was necessary. The indirect governor of the Jumo 004 engine was adapted to fit the BMW 003.

Production of the BMW 003 engine began in August 1944 in two versions:
- A-1 for the Arado Ar 234 C
- E-1 for the He 162

The two versions differed only in location of the attachment points. Two later versions, the A-2 and E-2, offered improved reliability and used alternate materials in their construction, however, only a few examples were completed by the end of the war.

A total of about 450 BMW 003 engines were built by the end of the war.

When the war ended the following versions of the BMW 003 were in planning:

003 C

seven-stage compressor and longitudinally-spaced stator; 900 kg static thrust

003 D

eight-stage compressor and two-stage turbine; 1,100 to 1,150 kg of static thrust; dimensions and weight same as BMW 003 A-1

Another variant was the BMW 003 R. This was a compound engine consisting of a BMW P-3395 (RLM: 109-718) liquid fuel rocket motor mounted on a BMW 003 A-1 turbojet. The engine was conceived to provide interceptor fighters with improved acceleration and speed.

When development work started, BMW designed the P3304 engine as a parallel/alternative to the P3302. It was designed as a two-shaft power plant with contra-rotating axial compressor rotors. This engine was designated 109-002 by the RLM. The project was abandoned in 1942 because of serious problems in developing the contra-rotating compressor.

BMW was already giving consideration to a turboprop engine in 1941, since this type of engine offered better fuel consumption than pure turbojets and was therefore more suitable for long-range aircraft. Such a power plant was projected under the designation BMW 028, but work was halted in 1942.

By deleting the propeller drive and one turbine stage, BMW derived a turbojet engine, the BMW 018, from its turboprop project. The new power plant was supposed to develop 3,400 to 3,500 kg of static thrust. Design work was completed by the end of the war. One prototype was almost complete, but it was probably destroyed in a bombing raid on the Spandau factory. It was planned that the BMW 018 turbojet engine should power the production version of the Ju 287.

Daimler-Benz Jet Engines

At the time that jet engine research began in Germany, Daimler-Benz was fully occupied with the development and manufacture of its line of high-performance aero-engines, the DB 601, DB 603, and DB 605. Therefore, it did not begin work on jet engines until 1939, after Prof. Karl Leist came over from the DVL. In keeping with his experience and expert knowledge, Prof. Leist was initially assigned to work on exhaust-driven turbosuperchargers for the high-altitude engines then under development.

In 1941, however, Daimler-Benz received a contract to develop a power

Daimler-Benz DB 007 turbofan engine on the test stand.

plant, which was not only more powerful than the BMW 003 and Jumo 004 engines, but which also consumed significantly less fuel. The engine was to have a maximum diameter of 900 mm and a maximum length of 5 m.

Prof. Leist proposed a turbofan engine, which the RLM designated 109-007. The engine was supposed to develop 610 kg of thrust at 900 km/h at an altitude of 6,000 m, which was equivalent to 1,400 kg of static thrust.

Development work began in 1941, and the first bench runs were carried out in 1943. Designed engine speed was reached in autumn 1943.

A high compressor ratio (calculated ratio was 1 : 8) and a high gas temperature were chosen in order to achieve the desired low fuel consumption. On advice from the AVA Göttingen a compressor

was designed which consisted of two contra-rotating drums. The inner drum had nine stages. The outer drum had eight stages on the inside and three stages on the outside (the second cycle). The two drums were linked by a planetary gearbox. Maximum revolutions were 12,600 per minute and 6,200 per minute. The compressor thus had 17 stages, which was the number of actual contra-rotating vane rings. The engine's bypass ratio was 2.42. The blade rings of the forward stages were made of light metal.

The last stage of the compressor was made of steel. On the first experimental devices the vane rings were made from complete rings, whereas on later versions the blades were attached using dovetail roots. Four individual combustion chambers were planned at first, however, this later had to be increased to five. Turbine entry temperature was 1,100 °C, which made it necessary to provide cooling for the turbine blades. The designers turned to a method that had previously yielded good results with turbosuperchargers. The turbine blades were cooled using air drawn from the second cycle (30%). The turbine itself was designed as a single-stage affair. The engine's complicated twin-flow layout prolonged design and testing. Entirely new experimental installations and test stands had to be built, delaying development.

Finally, in 1943, the RLM halted work on the project, having concluded that it would be several years before the engine was ready for production. It was hoped that engines in the 1,100 to 1,200 kg thrust class could be produced by developing the existing BMW 003 and Jumo 004 power plants, while the simpler Heinkel HeS 011 engine would cover the 1,300 kg thrust class.

By the time development work was halted, the first experimental engine had already run for more than 150 hours. The engine had a diameter of 840 mm and was 4.65 m long.

Supersonic Research in Germany

In 1939 the DFS (German Research Institute for Gliding Flight) carried out gliding experiments at altitudes as high as 11,600 meters. Then, in 1941 the RLM issued a contract for a high-altitude reconnaissance aircraft which was to be designed as a rocket-powered glider.

In the beginning the work had a low priority and proceeded slowly. Then, in 1943, the RLM assigned the high-altitude reconnaissance aircraft a high priority level, and at the same time ordered that construction of the aircraft, henceforth designated the DFS 228, go ahead.

One of DFS' main problems was that the high-altitude reconnaissance aircraft required a pressurized cockpit; however, because of the type's glider design it would have to be constructed of light metal.

The entire forward fuselage was laid out as a pressurized cockpit and was the only metal component of the aircraft. It was a cylindrical body with hemispherically-shaped front and rear ends. The cockpit was double-walled, and between the inner and outer hulls there was a layer of insulation consisting of aluminum isinglass. In order to achieve the smallest possible fuselage cross-section the pilot was placed in a prone position. A bulged plexiglass dome was installed in the nose of the aircraft. The pressurized cockpit was pressure stable to an altitude of 8,000 meters; above that altitude the pilot would have to wear an oxygen mask. The entire airframe of the DFS 228 was of wooden construction. The fuselage center-section housed the skid boxes, the fuel tanks, a Walter rock-

et engine, and two infrared cameras made by the Zeiss company. The rocket engine's combustion chamber was located in the rear fuselage.

The DFS 228's proposed operational profile was as follows:
1. The aircraft would be towed to an altitude of 10,000 m by a specially-modified Do 217 K-3.
2. The two aircraft would separate, and the DFS 228's rocket engine would propel it to an altitude of 22,500 m.
3. Switching on the rocket engine as required, the aircraft would maintain this altitude for 45 minutes and carry out its photo reconnaissance mission.
4. Head home in a glide to an altitude of 12,000 m, covering a distance of 750 km.
5. From an altitude of 12,000 m the aircraft could cover an additional 315 km, however, enemy aircraft could be expected to be present in that altitude range. Therefore, remaining fuel would be used for evasive maneuvers.

Conventional methods would have been inadequate to save the pilot at the heights at which the DFS 228 was expected to operate, and therefore a new procedure was developed:
1. The entire pressurized capsule could be separated from the fuselage by means of four explosive bolts, which the pilot activated by means of a lever.
2. Stabilized by a small parachute, the capsule would fall nose first to an altitude of 4,000 m.
3. Controlled by a barometer, at that altitude the plexiglass bubble would be jettisoned and the pilot would be ejected from the aircraft with his cradle.

DFS 228 high-altitude aircraft.

4. Once at a safe distance from the capsule the cradle would automatically separate from the pilot, who would then open his parachute.

Twelve aircraft were completed by the end of the war, and these underwent extensive testing at the *Erprobungsstelle Rechlin* with satisfactory results.

In the summer of 1944 the RLM ordered the DFS to develop a single-seat experimental aircraft from the DFS 228 for operation at high altitudes in the supersonic speed range. The project designation was DFS 346. Propulsion would be provided by two Walter liquid-fuel rockets, which were more powerful than existing turbojet engines.

Because of its anticipated high fuel consumption, like its predecessor the DFS 228, the DFS 346 would be carried to an altitude of 10,000 meters. From there the aircraft would climb under its

DFS 228 Technical Description

Purpose:	high-altitude reconnaissance aircraft with rocket propulsion
Crew:	one pilot in prone position in pressurized cockpit which forms fuselage nose
Wing:	cantilever mid-wing monoplane. Wing built of wood. Single continuous main spar from wingtip to wingtip, with wooden ribs and plywood skinning. Two-part ailerons of wood with fabric covering, inner sections acting as landing flaps. Braking flaps on upper and lower wing surfaces.
Fuselage:	three-part fuselage with circular cross-section. Forward section pressurized cockpit of metal construction, center and aft sections wooden monocoque construction.
Tail surfaces:	conventional cantilever tail surfaces of wooden construction, fixed surfaces plywood skinning, fabric covered control surfaces. Adjustable horizontal stabilizer mounted high on vertical stabilizer.
Undercarriage:	broad retractable metal skid beneath the fuselage section plus tail skid
Power plant:	one Walter HWK 109-509 A-1 liquid-fuel rocket motor capable of producing from 100 to 1 600 kg of static thrust. Tanks for C- and T-Stoff in fuselage center-section.
Military equipment:	no armament. Two Zeiss infrared cameras.

own power to altitudes of over 20,000 meters and reach a speed of 2,270 km/h (in excess of Mach 2.0).

Development of the new jet aircraft required innovative solutions in such diverse fields as aerodynamics, power plants, and materials. Theoretical calculations and wind tunnel tests were often at variance with the results of test flights. Vital knowledge had to be gained of engine performance at high altitudes, as well as flight mechanics and dynamics at high speeds. It had been realized that wing sweepback delayed the onset of compressibility effects (shockwaves) in the transonic speed range, however, the results that were achieved did not live up to calculated predictions. It should be noted, however, that by this time even the RLM had realized that jet aircraft offered the most effective, if not the only solution to the problem of defending against air attack.

When design work on the DFS 346 began in August 1944, it was possible to adopt many features of the DFS 228, such as the pilot's prone position. One novel requirement was that of positioning the instruments where they would not obscure the pilot's view. Some could be read by means of mirrors, which did not require the pilot to move his head. The escape system designed for the DFS 228 was also used in the DFS 346. The pressure capsule underwent some rede-

sign, however. The forward section now consisted of a hemispherical plexiglass panel. There was also an outer transparent panel. Warm air was blown into the space between the two transparencies to prevent icing. The DFS 346 was also fitted with a nose-mounted pitot tube. Because it was intended to operate at supersonic speeds, the DFS 346 was designed with a swept wing and was of all-metal construction.

Design work was completed in November 1944. Since DFS had no experience in building all-metal aircraft, production of the DFS 346 was assigned to the Siebel Flugzeugwerke in Halle an der Saale. A general description appears on the preceding page.

Several prototypes were in an advanced state of construction when American troops reached Halle in April 1945. Even though the DFS 346 never surfaced in America, there are indications that the Americans took a "prototype of a new supersonic aircraft" and 23 former Siebel employees with them when they pulled out.

After Soviet troops occupied took over Halle, Siebel continued work on behalf of the Soviets. Along with other companies in the Soviet occupation zone, Siebel was given the status of a "Special Design Bureau" (Siebel = OKB 3). The aircraft was subsequently designated the Siebel 346 (Si 346). The Sovi-

DFS 346 Technical Description

Purpose:	supersonic experimental aircraft
Crew:	pilot in prone position in pressurized cockpit which formed the fuselage nose
Wing:	two-part swept wing (45°) of all-metal stressed skin construction. Flaps along entire trailing edge, ailerons in middle, hydraulically-actuated flaps inboard and small wingtip flaps acting as high speed ailerons or trim tabs.
Fuselage:	all-metal monocoque fuselage with circular cross-section in three parts. Forward section pressurized cockpit with glazed nose cap, center section to accommodate fuel tanks and landing skid, aft section to house engine and tail section bearer.
Tail surfaces:	conventional all-metal cantilever tail surfaces. All surfaces swept 45°. Horizontal stabilizer mounted atop fin in T arrangement.
Undercarriage:	broad retractable landing skid
Power plant:	two Walter HWK 109-509 B-1 liquid-fuel rocket motors each capable of producing 2 000 kg of static thrust. Three fuel tanks for C- and T-Stoff in the fuselage center-section.

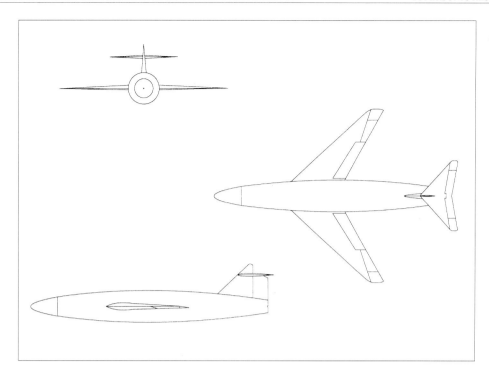

DFS 346 supersonic research aircraft.

ets were mainly interested in the type's aerodynamics and its Walter rocket engine.

Beginning on 22 October 1946 all special design bureaus in the occupation zone were dissolved and moved to Russian territory. The former OKB 3 was moved to Experimental Factory 1 near Podberezhye, approximately 120 km north of Moscow. With it went approximately 20% of the about 700 Siebel employees and their families. This factory housed an OKB 1 (Junkers) and an OKB 2, which consisted mainly of former Siebel workers. As well, an example of the DFS 346 which had been built at Halle was taken to the Soviet Union.

In Experimental Factory 1 the engineers succeeded in solving the problem of premature flow separation on the wings. After extensive wind tunnel experiments boundary layer fences were installed on the upper surfaces of the wings.

Testing was hampered by the absence of a suitable carrier aircraft, and there were also problems acquiring the special rocket fuels (*C*- and *T-Stoff*). In addition, important materials and items of equipment were not available. All of this was further complicated by the language problem and the need to convert all drawings and blueprints to Soviet standards.

Flight testing began in 1948 with gliding experiments. In 1949 a suitable carrier aircraft was found, one of three U.S. Army Air Force Boeing B-29s which had force-landed in Soviet territory during the war. These advanced bombers were copied by the Soviets, who produced their own version, which was dubbed the Tu 4. Later flight trails were conducted using a Tu 4.

June 1950 saw the completion of the Si 346 V3, which was fitted with a Soviet version of the Walter rocket engine. Initial flight trials were conducted without using the rocket engine. Powered test flights began in August 1951 and ended with the crash of the Si 346 V3 on 14 September 1951.

Thanks to the type's escape system, German test pilot Wolfgang Ziese, who tested all variants of the DFS 346 in the Soviet Union, was able to save his life, jettisoning the pressurized capsule and reaching denser air. After this loss the program was officially terminated. One of the main reasons for this was because, since 1947, the Soviet Air Ministry had been of the opinion that the German scientists and engineers were scientifically "exhausted" and should be sent home. In order to gain further use from these people the Soviets would have had to involve them in actual Soviet developments, something they did not wish to do.

Foreign Developments

Outside Germany, the nation which had made the greatest progress in the area of jet engines and their use as propulsion systems for aircraft was undoubtedly Great Britain.

In Italy the Campini Caproni N.1 had beaten the Gloster E28/39 into the air, flying on 27 August 1940, but results were so poor that development ceased in 1942. In the beginning the USA had no influence on the development of jet engines and jet aircraft. In September 1941 the USA received an English D.H. Halford H-1 turbojet engine, and from it developed the General Electric I-16 engine producing 748 kg of static thrust. Powered by this engine, the first American jet fighter, the Bell XP-59A, made its maiden flight on 1 October 1942. The Bell fighter's performance was poor, though, and was easily surpassed by contemporary piston-engined fighters. The Pentagon procured just fifty Bell P-59 Airacomets, which were used for test and training purposes.

In the Soviet Union, engine manufacturer A.L. Lyulka began development work on jet engines in 1936. He completed the design of a turbojet engine which was supposed to commence bench runs at the end of 1941, however, the German invasion put an end to these plans.

Consequently, from this point on my account will deal only with the British power plants which eventually powered the first British jet fighter, the Gloster Meteor.

The Englishman Frank Whittle began working on a gas turbine for aircraft in 1928. He presented his design to the British Air Ministry in 1929, but it was rejected. The aviation industry also showed no interest in his work, as it posed a great financial risk in their eyes.

After numerous attempts Whittle found a financial backer, and in 1936 he founded the company Power Jets Ltd. From then on things moved more quickly. The first combustion chamber experiments were carried out in October 1936, and the first experimental turbine, the Whittle U-1, was completed in March 1937. It was designed with an annular compressor. During one of the first test runs the compressor wheel failed. For financial reasons Whittle was forced to rebuild the turbine. On 12 April 1937 the turbine completed a successful test run. Engine speed could not be regulated, however, the run could still be judged successful. Following extensive technical modifications, at the end of 1937 the turbine reached a speed of 13,600 rpm. For the following turbine, the Whittle U-2, Whittle designed and built new turbine blades. In October 1938 the Whittle U-2 successfully completed a one-hour endurance run at 8,200 rpm, followed soon afterwards by a 105-minute run, developing 218 kg of static thrust.

Building on these results, Whittle constructed the U-3 turbine, which ran for the first time on 26 October 1938 and reached a speed of 16,000 rpm on 26 June 1939. The interest of the British Air Ministry was awakened.

In July 1939 the Air Ministry gave Whittle a contract to construct a turbine which would be suitable to propel an aircraft. At the same time Gloster received a contract to develop an aircraft, designated the E28/39, to accept the jet power plant. Gloster's internal designation for the aircraft was G.40, and it was completed in April 1940. It was a relatively small all-metal low wing monoplane, with a circular fuselage which tapered sharply towards the rear. The engine was installed in the fuselage center section aft of the cockpit. The air intake was located in the center of the nose, and the jet nozzle was not fitted with a thrust cone. The aircraft had trapezoidal wings with no landing flaps. On the other hand, the aircraft did have a pressurized cockpit and a tricycle undercarriage. Unlike their German counterparts, the British engineers were willing to try something new.

Because of delays with the Whittle W.1, the G.40 was not able to begin taxiing trials until 7 April 1941. Equipped with the W-1x experimental turbine, the aircraft was only able to reach a speed of 32 km/h on the grass surface. It was a disappointing result.

Meanwhile, on 12 April 1941 the Whittle W.1 turbine (385 kg static thrust at 15,000 rpm) was given flight clearance. The Gloster G.40 was immediately converted to take the Whittle W.1. The first taxiing trials took place on 14 May 1941. On that day the G.40 made three 200-meter hops. The next day, 15 May 1941, the Gloster G.40 made its maiden flight, powered by the Whittle W.1 turbine. The aircraft reached a speed of 600 km/h at an altitude of 6,100 meters, somewhat better than the contemporary Spitfire. At this point the British were firmly convinced that they were the first to successfully fly a jet aircraft, as was the case with the Americans on 2 October 1942.

Rolls-Royce became interested in jet engines in 1940. At first Rolls-Royce served as a subcontractor for Power Jets Ltd., acquiring the necessary know-how in this way. It is not surprising, therefore, that from 1944 Rolls-Royce took over development of the Whittle W.1 turbojet as the W-2B/23 Welland. This turbojet developed 771 kg of static thrust.

On 5 March 1943 the Gloster G.41 Meteor made its first flight powered by two Halford H-1 turbojets. It was the first jet aircraft to enter production outside Germany. The second British jet fighter, the de Havilland Vampire, took to the air for the first time on 20 September 1943, also powered by the Halford H-1 turbojet. The Vampire was not built in quantity until after the war. In July 1944 No. 616 Squadron became operational with sixteen Meteor Mk. 1 fighters. It was used in a defensive role, intercepting German V-1 flying bombs. The Meteor Mk. 3, powered by Rolls-Royce Derwent engines (907 kg static thrust), became available in January 1945. This variant was capable of reaching 800 km/h. Incidentally, the first production Gloster Meteor went to the USA in exchange for a Bell P-59 Airacomet.

(A completely different chapter concerns the advantages which the aviation industries of the Allied nations gained from their postwar examinations of German investigations into swept wings.)

While visiting Switzerland, Swedish engineer Lars Brising, later project director for the SAAB J29, was able to photocopy German research reports on swept wings. With the help of former Messerschmitt statistician Behrbohm, now an employee of SAAB, Brising was able to evaluate the data and put it to use. The result was the SAAB J29 *Tunnan*, which was based on the Messerschmitt P1101 concept. The Swedish fighter first flew on 1 September 1948.

The wing had "just" 25 degrees of sweepback, as the designers were reluctant to use more. More than 600 examples of the J29 were built, and the type served for many years, the last one flying until 1976.

The first American swept wing jet fighter was the North American F-86 Sabre (first flight 1 October 1947). In spring 1945 North American project engineers reviewed German swept-wing research data and Messerschmitt project work. They very quickly redesigned their jet fighter with a 35° swept wing similar in profile to that of the Me 262, with a profile thickness of 11%. According to these engineers, this saved them approximately three years of research. After examining documents and models in Göttingen and Oberammergau, George Schairer, development engineer on Boeing's B-47 project, redesigned the aircraft with swept wings. The first prototype flew on 17 December 1947.

Soviet engineers found documents concerning swept wings at the DVL in Berlin-Adlershof. This resulted in the first Soviet swept-wing aircraft, the Lavochkin La-160 research aircraft. It was based on the concept of the P1101 and had 35° of sweepback. The La-160 flew for the first time on 24 June 1947. Knowledge derived from it was used in designing the MiG 15 swept-wing fighter. The MiG 15 first flew on 30 December 1947, and the aircraft was built in large numbers.

After the war the western Allies banned German aviation specialists from working in their field for five years. At the same time many aviation specialists were employed by the French and Americans for periods of several months to several years. Both states had a great interest in the specialist knowledge possessed by these people, especially those involved with jet engines. As a result of the 1940 cease-fire, the French aviation industry had a lot of catching up to do, and it employed all possible means to do so.

Beginning in November 1945 Fritz Nallinger, formerly responsible for supercharger development with Daimler-Benz, assembled a team of engineers in Bregenz. Most were former Daimler-Benz workers, several of whom had worked on the DB 007 and DB/HeS 021 engines.

This team (125 engineers and technicians) was acquired by Turboméca. The team arrived in Pau in the Pyrenees in June 1946 and set to work on an engine in the 6,000 kg static thrust class. At the same time Turboméca projected a transoceanic aircraft to be powered by six of these engines. This plan was abandoned in 1947, however, and most of the Germans returned to Germany.

A small group did remain with the company, however, achieving leading positions, and it was ultimately partly responsible for the company's success in the small engine field, for example the Larzac engine of the Alphajet. It was this group, too, which developed the rotating combustion chamber which characterizes Turboméca engines to the present day. American troops occupied the industrial town of Stassfurt (south of Magdeburg) in the spring of 1945, and before handing it to the Soviets they moved the BMW team under Dr. Oestrich to the BMW factory in Munich-Milbertshofen. There he received offers from both the French and the Americans. He decided in favor of France. Immediately afterwards, Group "O" (Oestrich) was formed in a former Dornier branch plant in Lindau-Rickenbach in the French occupation zone. Over time, engine specialists from Junkers and Heinkel, as well as aircraft designers and aerodynamicists joined the group. In October 1945 Group "O" began development work on the ATAR 101 turbojet (ATAR = Atelier Aéronautique). In the summer of 1946 the group moved to Decize in France and was bolstered by numerous French specialists. In October 1945 the group was incorporated into SNECMA.

The ATAR 101 was designed to produce 1,700 kg of static thrust, however, the second prototype achieved 2,200 kg of thrust on the test stand. This engine, which formed the basis of many SNECMA power plants, was developed entirely by the German specialists of Group "O."

Dr. Oestrich became a French citizen on 1948 and was technical director of SNECMA from 1950 until he left the company in 1960.

In addition to the well-known rocket scientists, engine specialists were also taken to the USA as part of "Operation Paperclip." They were initially put to work in the U.S. Air Force research center in Dayton, Ohio. After completing their contracts the majority of them went to work in the American aviation industry. A small group, which included Pabst von Ohain, remained in Dayton.

Helmut Schelp went to Garrett AiResearch and played a significant role in the development of the TPE 331 turboprop engine.

Bruno Bruckmann, former head of engine development with BMW, went to General Electric, where he first assumed leadership of the J47 engine project. Later he was placed in charge of making the J79 engine production-ready. Bruckmann then headed development of the J79 engine, which was to power the North American XB-70 Valkyrie, until 1962.

In 1951 Dr. Anselm Franz joined Avco Lycoming, where he assembled several colleagues from his former Junkers team around him. There he developed the T53 gas turbine for the Bell UH-1D and the T55 for the Boeing Chinook, as well as their turbofan derivative, the ALF-502. He also played a significant role in development of the AGT 1500 gas turbine for the M1 Abrams tank. When he retired in 1968 Dr. Franz held the positions of vice-president and assistant general manager.

The Soviet Union took a very different approach. After occupying the industrial facilities in Stassfurt, they set up the OKB-2 in the former BMW factory there. Its main task was to assemble BMW 003 engines from available components. The first bench runs were carried out at the beginning of 1946. When the war ended the Soviets were aware of the plans to increase the output of the BMW 003 and Jumo 004 engines, and they therefore demanded that the BMW engine's output be raised to 1,100 kg of static thrust. In the summer of 1946 a test engine completed a 50-hour endurance run in which it developed 1,000 kg of static thrust. The program was canceled and efforts switched to copying and building the BMW 018 engine.

On Soviet orders the Junkers factories in Dessau built several Jumo 004 engines for test purposes. The Soviets then ordered Junkers to make the Jumo 012 engine, which was in the project phase when the war ended, ready for production. Without warning, during the night of 21-22 October 1946 the Soviets began transporting the technical personnel, their families, and all installations and test benches of the BMW and Junkers factories to Upravlencheski Gorodok, 20 kilometers south of Kuibyshev on the Volga. The move affected 250 specialists from BMW and 350 from Junkers.

Work continued in the Soviet Union. There the Junkers group developed an engine in the 3,000-kg thrust class based on the Jumo 012. After repeated blade failures the program was halted in

1948. The BMW and Junkers groups were subsequently combined to continue development work on the Jumo 022 turboprop project.

In 1950 the engine completed a 100-hour run and was certified for production as the Kuznetsov NK-4. The engine subsequently powered the AN-8, AN-10, and IL-18.

Most of the specialists and their families were returned to Germany in 1953. The remaining German specialists were concentrated in Savyolovo (100 km north of Moscow) where they were given the task of designing the airframe and engines for a civil aircraft project. The design was supposed to be realized in the GDR. The group developed the 3,300-kg thrust Pirna 014 engine, which was based on the Jumo 012. Powered by four of these engines, the 152/II commercial aircraft completed its maiden flight on 26 August 1960. The type did not enter production.

The postwar period was quite different for pilots. Soon after the war Prof. Kurt Tank and several Focke-Wulf engineers went to Argentina, where they developed the *Pulqui II* fighter, which was based on the Focke-Wulf Ta 183 *Huckebein* project. The fighter, which had 40 degrees of wing sweepback, made its first flight on 16 June 1950.

Prof. Willi Messerschmitt went to Spain, where he developed the HA 200 jet trainer and the HA 300 interceptor fighter.

Appendix

Technical Overview

Name	Type	Crew	Wingspan (m)	Length (m)	Height	Power Plant	Output (kp)	Max. Speed (km/h)	at altitude of (m)	Empty Weight (kg)	All-up Weight (kg)	Armament (kg)
He 178	E	1	7.20	7.48	–	1 HeS 3 B	500	700	–	1 560	1 998	none
He 280	J	1	12.20	10.40	3.06	2 HeS 8 A	2 x 700	776	6 100	3 055	4 215	3 x MG 151/20
He 162 A-1	J	1	7.20	9.05	2.60	1 BMW 003 E	800	840	4 000	1 758	2 805	2 x MG 151/20
He 343	B	2	18.00	16.50	5.35	4 HeS 011	4 x 1 300	910[1]	–	10 770	19 550	2 x MG 151/20; 3 000 kg of bombs
Me 262 A-1a	J	1	12.50	10.60	3.83	2 Jumo 004 B-1	2 x 900	873	6 000	4 000	6 775	4 x MK 108
Me 262 A-2a	Jabo	1	12.50	10.60	3.83	2 Jumo 004 B-1	2 x 900	750	6 000	4 000	7 100	2 x MK 108; 2 x SC 250 bombs
Me 262 B-1a/U1	NJ	2	12.50	10.60	3.83	2 Jumo 004 B-1	2 x 900	795	6 000	4 764	7 600	4 x MK 108
Me 328	J	1	6.90	7.17	1.60	1 Argus As014	360	805	–	1 600	4 500	500 kg explosive charge
Me P1101	E	1	8.08	8.83	–	1 Jumo 004 B	900	1 100[1]	6 000	2 077	4 070	none
Ar 234 B-2	B, A	1	14.41	12.65	4.42	2 Jumo 004 B	2 x 900	780	6 000	5 211	9 350	2 x MG 151/20; 1 x 1 000 kg bomb or 3 x 500 kg bombs
Ar 234 C-3	B, A, NJ	1	14.41	12.65	4.42	4 BMW 003 A-1	4 x 800	890	6 000	5 850	11 555	2 x MG 151/20; 1 x 1 000 kg bomb + 2 x 250 kg or 3 x 500 kg bombs
Ju 287	B	2	20.10	18.28	–	4 Jumo 004 B-1	4 x 900	814	3 000	–	22 250	none
Ho IX V3		1	16.80	7.47	2.81	2 Jumo 004 B-1	977[1]	–	5 067	5 067	7 726	none
Hs 132 A	S	1	7.20	8.90	–	1 BMW 003 A	800	780	4 000	–	3 400	1 x 1 000 kg bomb
Li P13a	J	1	6.00	6.70	3.25	ramjet propulsion	–	–	–	–	2 300	2 x MK 108
Ta 183	J	1	9.50	8.90	–	1 HeS 011 A	1 300	955	7 000	2 830	4 200	2 (4) x MK 108
DFS 346	E	1	8.98	11.65	3.50	2 HWK 109-509 B-1	2 x 2 000	2 270[1]	20 000	–	5 200	none

Legend:

A	reconnaissance aircraft
B	bomber
J	fighter
S	close-support aircraft
E	experimental aircraft
Jabo	fighter-bomber
NJ	night fighter
1	theoretical

Chronology of Jet Engine and Aircraft Development to 1945

Power Plants

1937	March 1937	Heinkel HeS 02 turbine runs for the first time
	12/04/1937	Whittle U-1 turbine runs for the first time
1938		
	spring 1938	Heinkel HeS 03A turbine runs for the first time
1939		The RLM issues a requirement for a turbojet capable of producing 600 kg of thrust
1940	11/10/1940	Heinkel HeS 08 engine runs for the first time
1941	20/02/1941	BMW P3302 engine runs for the first time
1942	15/03/1942	First flight of the Jumo 004 on an Me 210
1943	January 1943	Production of the Jumo 004 begins
	May 1943	Jumo 004 B runs for the first time
1944	early 1944	BMW 003 A-0 flight tested on Ju 88
	January 1944	Production of Jumo 004 B-1 begins
	August 1944	Production of BMW 003 A-1 and BMW 003 E-1 begins
1945		

		Aircraft
1937		
1938		
1939	27/08/1939	Maiden flight of He 178 with one HeS 03B, first aircraft in the world to fly under jet propulsion
1940	27/08/1940	First flight of the Campini Caproni N1
1941	02/04/1941	Maiden flight of the Heinkel He 280 powered by two HeS 08, world's first twin-engined jet aircraft
	15/05/1941	Maiden flight of the Gloster E 28/39
1942	18/07/1942	Maiden flight of the Me 262 V3 powered by two Jumo 004
	02/10/1942	Maiden flight of the Bell XP-59A
1943	05/03/1943	Maiden flight of the Gloster Meteor
	15/06/1943	Maiden flight of the Ar 234 V1 powered by two Jumo 004
	20/09/1943	Maiden flight of the de Havilland Vampire
	09/12/1943	Formation of Erp.Kdo. Thierfelder, Me 262 service trials unit
1944	08/01/1944	Maiden flight of the Lockheed P-80 Shooting Star
	01/02/1944	Maiden flight of the Ar 234 V8 powered by four BMW 003s, the first four-engined jet aircraft in the world
	28/03/1944	Maiden flight of the Me 262 S21 pre-production aircraft
	19/04/1944	First Me 262 enters service with Erp.Kdo. Thierfelder
	20/06/1944	Formation of Erp.Kdo. Schenk, Ar 234 service trials unit
	16/08/1944	Maiden flight of the Ju 287 V1 powered by four Jumo 004s
	27/10/1944	Formation of III./EJG 2, first jet unit in the world
	06/12/1944	Maiden flight of the He 162 powered by one BMW 003
1945	02/02/1945	Maiden flight of the Horten Ho IX powered by two Jumo 004s

Umrechnungstabelle

1 kp	0,0098 kN
1 PS	0,7355 kW
1 ft	0,305 m
1 nm	1,852 km
1 kg/kph	101,972 kg/kNh

Abbreviations

AVA	Aerodynamic Research Institute
Bramo	Brandenburgische Motorenwerke
C-Stoff	rocket fuel (30% hydrazine hydrate, 57% methanol, 13% water, remaining components calcium-copper-cyanide and hypergol with H_2O_2)
DFS	German Research Institute for Gliding Flight
DVL	German Aviation Research Institute
E-Stelle	Test Station
FHL	remotely-controlled tail barbette
FuG	radio device
LFW	Vienna Aviation Research Institute
MG	machine-gun
MK	machine-cannon
PTL	turboprop engine
TL	turbojet engine
T-Stoff	rocket fuel (80% hydrogen peroxide)
ZTL	turbofan engine

Bibliography

Brown, Eric: *Berühmte Flugzeuge der Luftwaffe 1939-1945*, Motorbuch Verlag

Brütting, Georg: *Das Buch der deutsche Fluggeschichte*, Drei Brunnen Verlag GmbH, 1979

Ciesla, Burghard: *Top Secret DFS 346 Überschall-Forschungsprojekt aus Deutschland*, Flugrevue 12/97

Dabrowski, Hans-Peter: *Deutsche Nurflügel bis 1945*, Podzun-Pallas Verlag 1995

Dabrowski, Hans-Peter: *Fliegendes Dreieck, Lippischs Triebflügel*, Flugrevue 09/98

Donald, David: *Airplanes of the Luftwaffe*, Aerospace Publishing London, 1994

Dressel, Joachim and Griehl, Manfred: *Taktische Militärflugzeuge in Deutschland 1925 bis heute*, Podzun-Pallas Verlag 1992

Ebert, Hans J: *Willy Messerschmitt—Pionier der Luftfahrt und des Leichtbaues, Die deutsche Luftfahrt Bd. 17*, Bernard & Graefe Verlag, 1992

Gersdorff, Kyrill von and Grasmann, Kurt: *Flugmotoren und Strahltriebwerke: Die deutsche Luftfahrt Bd. 2*, Bernhard & Graefe Verlag, 1985

Green, William: *Famous Bombers of the Second World War*, MacDonald Publishers, London, 1967

Johnson, B.: *Streng Geheim*, Paul Pietsch Verlage GmbH, Stuttgart

Jurleit, Manfred: *Strahljäger—Me 262*, Motorbuch Verlag, 1995

Lange, Bruno: *Das Buch der deutschen Luftfahrttechnik*, Verlag Dieter Hoffmann, 1970

Nowarra, Heinz J: *Die deutsche Luftrüstung 1933-1945*, Bernard & Graefe Verlag, 1993

Radinger W. and Schick W.: *Me 262*, Aviatic Verlag, 1996

Smith, J. Richard and Creek, Eddie: *Jet Planes of the Third Reich*, Monogram Aviation Publications, Boylston, Massachusetts, 1982

Ziegler, Mano: *Kampf um Mach 1*, Ehapa-Verlag 1965

Wehrtechnische Studiensammlung – BWB Koblenz.

Photo Credits

My thanks to the following persons and institution for allowing
the use of illustrations:

Herr Lothar Nebgen
Herr Willy Radinger
Deutsches Museum, Munich
Wehrtechnisches Studiensammlung – BWB Koblenz
DaimlerChrysler Aerospace
MTU Motoren- und Turbinenunion GmbH

Also from the publisher

The Luftwaffe Profile Series Number 1: Messerschmitt Me 262. Manfred Griehl. Number 1 in the Luftwaffe Profile Series describes the design and use of the Messerschmitt Me 262.
Size: 8 1/2" x 11" ■ b/w and color photographs, color profiles, drawings ■ 52 pp.
0-88740-820-6 ■ soft cover ■ $14.95

The Luftwaffe Profile Series Number 8: Arado Ar 240. Gerhard Lang. Number 8 in the Luftwaffe Profile Series describes the design and use of the Arado Ar 240.
Size: 8 1/2" x 11" ■ b/w photos color profiles, drawings ■ 24 pp.
0-88740-923-7 ■ soft cover ■ $9.95

The History of German Aviation: Kurt Tank - Focke-Wulf's Designer and Test Pilot. Wolfgang Wagner. This volume, the second in a mulit-volume history of German aviation, discusses the life work of one of Germany's greatest aircraft designers in pictures, tables, drawings and in his own words discussing flight testing of his airplanes. Examined are the birth of Focke-Wulf's airplanes, the ideas and thinking which formed the foundation of Tank's designs, his masterpieces in the areas of long-range and high-speed flight as well as flight test results. The reader experiences the metamorphosis of an aircraft design from the first pencil line on the drawing board to the screaming, nearly supersonic dives during its evaluation phase. Tank would only entrust to civilian pilots, military flyers and the captains of the Lufthansa fleet those designs whose superior flying characteristics, stability, and flawless stall handling he had been able to experience first hand. A mixture of technical history, flight testing and previously unpublished data enable the reader to catch a fascinating glimpse of the aircraft built during the 1930s and 1940s, not to mention the outstanding designs Tank worked on in other countries following the war. All of the classic Focke-Wulf aircraft are to be found: the well-known Fw 200, Fw 190, Ta 152, and Ta 154, as well as the lesser known Fw 57, Fw 187, Fw 189 and others.
Size: 8 1/2" x 11" ■ over 200 b/w photographs and line art ■ 272 pp.
0-7643-0644-8 ■ hard cover ■ $39.95

The History of German Aviation: Willy Messerschmitt - Pioneer of Aviation Design. Ebert/Kaiser/Peters. Willy Messerschmitt (1898-1978) was indisputably one of the most significant of Germany's aeronautical design engineers. This book examines Messerschmitt's life as a designer, aircraft builder, and businessman; it begins with the Harth-Messerschmitt gliders (1913-1923), winds its way through the sportplanes of the 1920s and 1930s, the M 18 and M 20 passenger airliners, the Bf 108 Taifun commuter plane, the Bf/Me 109, 110, 210, 410 military aircraft, and continues on to the Me 261, 264, 321, and 323 behemoths, as well as the Me 262 jet powered combat plane and swing-wing P.1101. The activities of the Messerschmitt AG after World War II, with the forced interruption of German aircraft production and Professor Messerschmitt's foreign developments in Spain and Egypt, are also given a thorough treatment.
Size: 8 1/2" x 11" ■ over 670 b/w and color photographs, line drawings ■ 416 pp.
0-7643-0727-4 ■ hard cover ■ $49.95

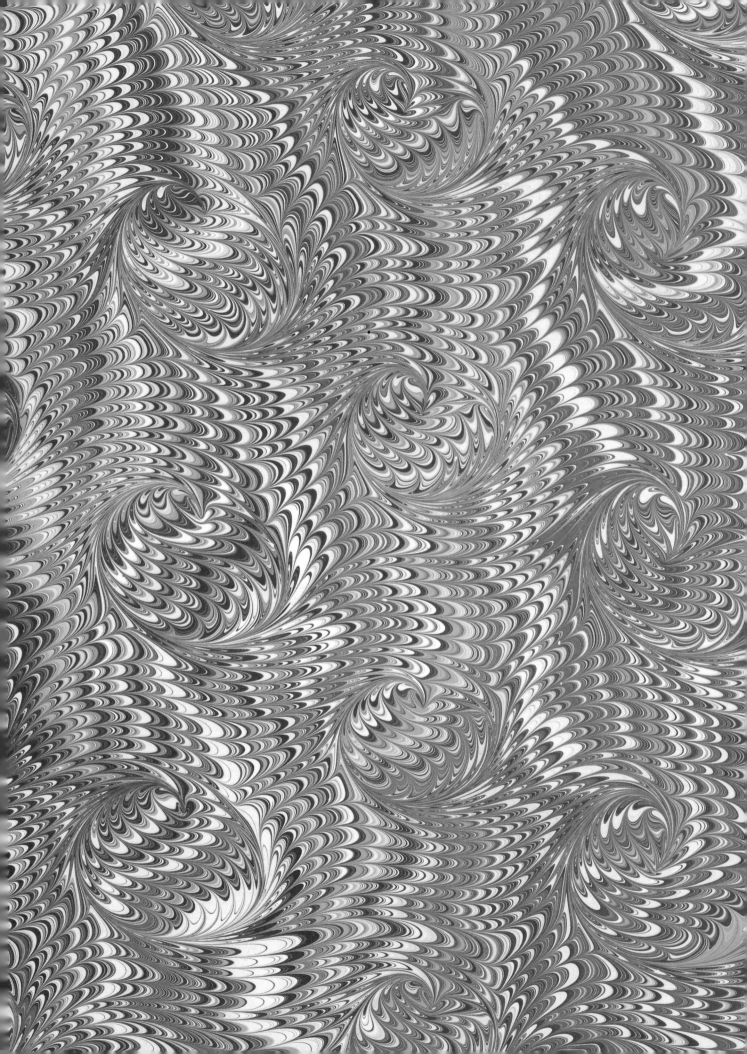